MARGARET BEAUFORT
LIBRARY
INSTITUTE

QH325 .P75
8F277
Ø88 824 Ø71

The Origin and Evolution of Life

THE BRIDGE SERIES

---

editor

F. C. BROWN M.A., B.SC., F.R.I.C.
Head of Department of Physical Sciences,
University of London Institute of Education

---

energy and bonding

MICHAEL HUDSON PH.D., B.SC., A.R.I.C.

the mechanical properties of matter

M. T. SPRACKLING PH.D., A.INST.P.

JOHN T. PRICE M.A. (Cantab.)
Head of Biology Department,
Bishop Wordsworth's School, Salisbury

the origin and evolution of life

THE ENGLISH UNIVERSITIES PRESS LTD
St Paul's House Warwick Lane London EC4

ISBN 0 340 05074 8

First printed 1971

Printed and bound in Great Britain for The English Universities Press Ltd
by T. & A. Constable Ltd., Edinburgh

editor's foreword

The transition from school to university or college is, for many students, a difficult process and this is reflected in the significant number of failures occurring in the first years of university courses. Contributory causes include a significant increase in pace, drastic changes in teaching and learning patterns and an increased emphasis on private study and book-usage.

The purpose of this new series of texts is to bridge the gap between school and university or college by supplying concise and up-to-date presentations of important topics in the broad spectrum of science, mathematics and engineering. Some of the topics will be ones less frequently encountered in the sixth form curriculum, others will extend and enrich more familiar themes. The books will concentrate on essential principles, with worked-out problems where appropriate.

The texts should provide valuable stimulus to the able sixth former, prior to leaving school, while they should be of the greatest assistance to the first year student in university, college of education or college of further education.

F.C.B.

preface

In accordance with the general aim of the series, this book is an attempt to present an account of what is known or what is postulated about the origin of life, and the subsequent evolution of the main animal and plant groups. The subject is one of prime interest to many—not only to biologists—and, in contrast to other books in the series, *The Origin and Evolution of Life* cannot be described purely as a textbook. Certain basic scientific facts are included, it is true, but the majority of the contents consist of hypotheses which are never likely to be experimentally verified; and the reader should treat the book as a source of facts, ideas, and theories on which thought and discussion can be based.

The aim throughout has been to emphasise the ecological pressures which have resulted in evolutionary advance, although where these are not clear, the main structural changes that have occurred in a group are described factually.

I have drawn on many sources for this book, and indeed, one of my objectives in producing the volume was to gather together the many ideas about the course of evolution that are scattered so widely throughout journals, scientific textbooks, and philosophical treatises.

I should finally like to thank my wife for carefully interpreting my rough sketches and producing the line drawings which illustrate the text.

Salisbury 1970 John T. Price

contents

1
the origin of life

What is life? What is man's place in nature and the universe? Where, when and why did life originate? These are questions that men have asked themselves and each other since the earliest civilisations of which we have records. Certain scientifically deduced facts now help to provide a lead to the answers of some of these questions, and the purpose of this book is to attempt objectively to analyse and solve these problems to the best of our present-day ability.

The Main Theories

The various theories that have been proposed can all be considered as variants of four main hypotheses:

1. Life has no origin. Life, matter and energy are co-existent in an infinite and eternal universe. According to this theory, "seeds" of life travel from one planet to another through space.
2. Life was created by a supernatural event at a particular instant of time in the past.
3. Life arrived on this planet from elsewhere in the solar system or universe.
4. Life arose on this planet strictly according to chemical and physical laws as we know them.

Of these theories, 1 is not consistent with the facts as they are known today, and 2 is not capable of scientific investigation and is therefore outside the scope of this discussion. Theory 3 is unsatisfactory in that it merely refers the problem to some other point in space and time, and unlikely because of the almost insuperable difficulties involved in crossing interstellar or interplanetary space. These difficulties include the completely anaerobic conditions, which would result in dehydration and the necessity for a complete suspension of all metabolic processes, and the more serious effects of the raw radiation met in space, which would almost certainly prove lethal to any living organisation which was unprotected. This last criticism also applies to the "seeds" invoked in the first theory.

We are now left with the fourth theory, that life originated and evolved on this planet. It is the most satisfying theory to investigate, but it raises many more problems in the course of its investigation.

What is Life?

Life as we know it is easy to recognise and hard to define. We can look at the vast majority of organisms and immediately recognise them as possessing the property of "life". When it comes to definition, however, it is impossible to produce a concise statement as to what this "life" is. We talk in terms of processes such as growth, reproduction, respiration and irritability, none of which are the sole prerogatives of living organisms, and all of which have to be included if we are to be at all correct in the delimiting of our category "living organism". There are also the problem cases. Viruses, the dormant stages of many plants and animals may show none of our carefully listed properties, yet they are alive. Or are they? The problem here becomes entirely semantic, and is not worth the energy that is often wasted in discussing it. Biology is not a science of clear-cut definitions, and life itself cannot be so easily delimited.

It does help, however, to consider the million and a half species of organisms inhabiting the earth today and observe whether there are any characteristics that all have in common. If we can find some such features, it is at least reasonable to assume that these are fundamental to life as we know it, and furthermore that they were probably present in the earliest stages of the evolution of life.

Protein, Deoxyribonucleic Acid and Ribonucleic Acid

The most apparent common factor of living organisms is that proteins play a very important part in their structure. Deoxyribonucleic acid (DNA) and ribonucleic acid (RNA) are also universally present, either together or separately. In view of our present knowledge of the functioning of the genetic code, the fact that proteins and nucleic acids are so fundamental is not particularly surprising. The hypothesis of one-gene-one-enzyme (or other protein) is now fairly well established, and the coding mechanism of base triplets corresponding to particular amino-acids is now so well known and so often written about, that it is certainly not necessary to enter into any extra descriptive detail here. It must be remembered, however, that a nucleic acid of some type is essential in all living organisms to effect the processes of reproduction growth and differentiation.

Adenosine Diphosphate and Adenosine Triphosphate

The other striking feature that all living organims have in common, is the presence of an ADP-ATP system as the energy storing mechanism. This involves the synthesis of adenosine triphosphate from adenosine diphosphate and inorganic phosphate, in regions and at times when surplus energy is being released, and the breakdown of adenosine triphosphate to adenosine diphosphate and phosphate at some later time when energy is required. Many other high-energy bonds could theoretically be used similarly, and the universal occurrence of the ADP-ATP system strongly suggests that it is the energy storing mechanism that has been passed down from the very earliest forms of life.

The Qualities of Life

It will now pay to reconsider the question posed earlier as to the nature of life. This time, however, instead of looking for the common factors present in all living organisms today, let us consider the faculties or properties that must be possessed by the simplest system that we would call "living". There can be disagreement at this level, of course, but it would be generally held that the system must be capable of replicating itself: that is, producing an exact copy of itself, with no assistance from its environment beyond the provision of raw materials. The replica so produced must have the same ability as its parent, to replicate itself, and so on *ad infinitum*. Assuming that our living system is a complex one, composed of more than one basic unit, then it must have the ability to select from the environment, and hold, the units necessary for producing another of its kind. These selected units must then be joined together in the correct order, and the new replica released to lead its own existence. Another problem arises here, however, as the synthesis of large chemical complexes from smaller units always involves either energy uptake or, less frequently, energy release. In the former case, where is the extra energy to come from? One further point is that even in exergonic syntheses, the reaction will not proceed unless some catalyst is present, which will reduce the energy of activation required for the synthesis. Thus our hypothetical elementary living system must possess the following features:

1. The ability to produce a replica of itself from the raw materials in its environment,
2. A source of energy to effect the synthesis. This source could be external or internal,
3. A catalyst or catalysts to permit the synthesis to occur, or to control its rate.

It is reasonable to assume that the earliest form of life must have accorded with the criteria outlined here, but further speculation along these lines is fruitless, unless we know something of the conditions that must have prevailed on this planet when the events we are postulating occurred.

The Origin of the Earth

It is generally agreed by astronomers, geologists and biologists that the Earth is some 5,000 million years old, and that at the time of its formation the temperature of the planet was considerably higher than it is now. There is less agreement, however, as to the actual origin of the Earth. Some hold that it gradually formed from hot gases thrown off by the Sun, whilst others believe that many small solid objects, such as meteorites, gradually aggregated to form one large solid planet. What is absolutely certain, however, is that the Earth in those days bore very little resemblance to its present condition. The surface of the planet was, of course, absolutely barren, with a high level of volcanic activity, and continuous earth movements of all magnitudes, as the cooling crust contracted, folded and fractured.

The atmosphere of the Earth, too, must have been totally different in those early days. The temperature of the crust would for a long time have been above 100 degrees Celsius, and therefore all the water must have been present in the atmosphere as vapour, until the Earth cooled sufficiently to allow water to condense in pools at certain points on its surface. The volcanic activity must have produced large quantities of hydrogen, ammonia and methane; and some carbon dioxide and nitrogen, together with other hydrocarbons. This we can deduce from volcanic action today. Also there must have been a total absence of free oxygen, as it would have immediately combined with various materials that then existed in a chemically reduced state, as well as free iron, sulphur and other elements. The amount of hydrogen in the atmosphere must have gradually decreased as the Earth cooled, because the rate at which it was being released by volcanic activity must have eventually dropped below the rate at which it was being lost from the atmosphere on account of its low density. Despite this, however, there is no doubt that there was a reducing atmosphere present at the stage in the Earth's history when life first arose. This may seem most unlikely in view of the fact that almost all living organisms today are dependent on oxygen for respiration, but as Oparin first pointed out in 1923, the presence of a reducing atmosphere during the early stages of life was, seemingly paradoxically, a necessity.

Oparin was the first person to suggest that the atmosphere of the primitive Earth was not the same oxygen-nitrogen mixture that exists today. He argued from the evidence of other planetary atmospheres, most of which contain large quantities of hydrogen, methane, ammonia and carbon dioxide; and since then, several pieces of practical evidence, as well as numerous theoretical considerations, have strengthened his hypothesis almost to the point of certainty.

As so often happens in science, this radical reform of thought stimulated many other people into speculation, and some into action. Probably the most exciting and informative work has come from a series of experiments which were started by Stanley Miller in the USA. Attempts were made to simulate the conditions that were thought to have existed during the Earth's infancy, both as far as chemical and physical conditions were concerned. Mixtures of the gases methane, ammonia and carbon dioxide, together with water vapour, were subjected *in vitro* to high temperatures (80-90° C), light (including ultra-violet wavelengths), ionising radiation and electrical discharge. Subsequent analysis of the contents of these reaction chambers showed that small quantities of a great variety of organic substances had been produced. Many of these substances were, and are, of considerable biological importance, including many organic acids of fairly low molecular weight, several amino-acids including glycine and alanine, urea, adenine, and simple sugars such as ribose. More recent experiments of a similar nature have demonstrated the formation of simple proteins, ATP and porphyrins under the same reducing conditions.

A particularly interesting feature of organic substances is that where optically isomeric forms of a compound are equally likely to occur, in fact only one form occurs in living organisms. This is still a mystery, but it is very significant that in experimental work on heat polymerisation of polypeptides and other comparatively simple organic units, it has been found that either the pure d- or the pure l-form is more stable than a mixture of the two. The seemingly asymmetrical distribution of these optically active substances in living organisms may therefore be a result of only one pure isomer surviving at a very early stage in evolution.

It is thus very reasonable to assume that the type of syntheses carried out by Miller, must have occurred in the Earth's early stages, and the gradual condensation and accumulation of water on the surface of the planet would have provided a solvent for these compounds. The Earth would therefore have been covered with numerous puddles, ponds and lakes of a dilute organic "soup". Urey has estimated that the first seas may have been a 10 per cent solution of organic compounds. It is important to realise, however, that the proportion of dissolved inorganic salts must have been very small indeed, and the build-up to the present level of 3½ per cent must have occurred very slowly as salts were gradually leached from the rocks of the Earth's crust.

The First Signs of Life

It has therefore been shown, theoretically and practically, that given the conditions that are likely to have prevailed whilst the Earth was cooling down, a whole range of organic compounds would have been produced, most of which today are only found as metabolic products of living organisms. It is tempting at this stage to oversimplify the problem and equate the formation of these organic compounds with the origin of life. None of the products, however, is capable of replication, and although replication is the next step that must have occurred in the evolution of a "living" organism, it would be a great mistake to assume that this step was inevitable.

The great biochemical uniformity of the self-replicating mechanisms found in existing organisms, suggests that a nucleic acid was almost certainly the first, or at least the most efficient among the first, self-replicating system to arise in the chemical evolution of this planet. The composition of a nucleic acid is simple: a ribose sugar, phosphate groups and a limited number of comparatively simple nitrogenous bases. Five of these bases are found in nucleic acids today, but Crick has suggested that there might only have been two in the primal nucleic acid. All these components have been shown to be formed under the electric discharge conditions described above, but they do not, of course, automatically combine together in the right sequence to form a DNA molecule. It is true that recently RNA has been synthesised *in vitro* from its components, but only in the presence of the specific enzyme complex that catalyses the reaction. There are thus two major problems facing us in our attempts to suggest a mechanism whereby a self-replicating system could have arisen. The first is one of the catalysis of certain key reactions, and the second is one of concentration. It is difficult to see how, in a dilute soup of such varied composition the necessary components for the synthesis of say, RNA, could come together with the required catalyst, unless they were somehow aggregated.

Catalysis

As has been already pointed out, the catalysis of biological reactions is carried out almost exclusively by enzymes, which consist usually of a protein base plus a prosthetic group. This prosthetic group is sometimes a metallic ion and sometimes a vitamin. Enzymes are therefore comparatively complex organic substances, and are produced in living organisms today by the coding and synthetic activities of the nucleic acids. We thus have a chicken-egg situation: enzymes can only be produced in the presence of nucleic acids, and the synthesis of nucleic acids is dependent on the catalysis of an enzyme. It is possible to resolve this paradox, however, and the clue to the explanation lies in the prosthetic group of the enzyme. Many metallic ions are catalysts of both inorganic and organic reactions. They bring about organic reactions much more slowly than do the specialist enzymes, but they do enable certain reactions to proceed to completion which otherwise would not take place at all. It is thus at least possible that the synthesis of the first nucleic acid was catalysed by simple ions of copper, iron or vanadium, and that these inorganic ions may also have catalysed numerous other organic syntheses. These ions may then have combined with porphyrins or polypeptides of abiogenic origin, and certain of these newly formed compounds may have been of far greater efficiency in catalysing a particular reaction than the ion itself. Thus natural selection acting at this early stage may have selected the ion-protein complex or enzyme in preference to the simple ionic precursor. This does assume that individual "organisms" had arisen by this stage, on which the natural selection could act, but as we shall see shortly, this is not at all unlikely. This suggested outline is entirely hypothetical, of course, but the presence of the more highly catalytic metallic ions as prosthetic groups in present day enzymes does give a certain amount of credibility to the hypothesis.

Concentration

As we have remarked above, the coming together of the various organic and inorganic components required for the synthesis of a nucleic acid must have been a very rare occurrence in the Earth's first seas. A well-known property of colloidal solutions, however, has been suggested as the means by which concentration and comparative stability were first attained. This phenomenon is known as coacervation. If a colloidal solution of similar composition to the Earth's early seas is produced, it shows a marked tendency to split up into droplets suspended in a continuous aqueous phase. These droplets are rich in the colloidal solutes, and the surrounding medium, with which the droplets are in equilibrium, is relatively free of the colloids. Crystalloids also tend to be drawn into these droplets with the colloids. As Oparin remarks: "It is hardly possible to find any other equally effective means for concentrating protein-like substances and others of high molecular weight, especially at low temperatures." Following on this concentration process, chemical reactions would occur inside the droplets, and some of the products of these reactions would make the coacervates unstable. These droplets would then break down and disperse. More complex organic compounds would have been constantly produced, and to maintain the equilibrium, the coacervate drops would have absorbed more organic solutes as they became available. Thus the drops would grow bigger, until, reaching their optimum size, they would break up into two or more smaller droplets. We thus have a system whereby the organic compounds first produced could have been concentrated, and also a system which would have perpetuated the more stable coacervate systems. One further very significant fact about coacervate drops, is that lipid molecules in the solution tend to align themselves along the boundary between the drops and the medium, producing a stabilising effect on the drop itself. This is important because the cell-membrane is composed largely of orientated lipid molecules, and it is possible to see rather more easily how the coacervate drop might have evolved into the first living acellular protistan.

It is important to note at this stage that the ultra-violet radiation, which undoubtedly aided the synthesis of the simpler organic compounds, in fact tends to break down the more complex polymerised organic compounds. Therefore the formation of nucleic acids and polypeptides must have taken place out of the direct irradiation of ultra-violet wavelengths, probably at a depth of 9-12 metres of water, which would provide the necessary shielding effect. This fact also renders it unlikely that the first self-replicating chemical compounds arose on the shores of the early seas, as has been suggested by some workers.

One other suggestion which has recently been put forward is that life originated on land rather than in water. This intriguing theory is based on observations of living organisms that can be totally dehydrated and left to remain in a state of suspended animation until they are once more placed in water when they resume their normal life processes, apparently unharmed. This theory proposes that concentration of the primal organic compounds occurred in small splash puddles and niches on land. Evaporation would provide the concentration required for more complex syntheses, and when dehydration was complete, the desiccated simple organic substances could remain *in situ*, or be blown to other niches until rehydrated later by a solution containing other compounds synthesised at a later date which may have been necessary for the crucial polymerisations. This is quite an acceptable alternative theory, the only drawback being that the degree of exposure to the ultra-violet radiation would make it seem unlikely that polymers above a certain level of complexity could have arisen.

The Changing Atmosphere

It has already been emphasised how different the Earth's early atmosphere was from our present day one, and how it was necessary for reducing conditions to be present in order for the first polymerised organic compounds to form. We must now consider how the changes occurred which brought about the alteration of an atmosphere completely hostile to life as we know it, into our present oxygen-rich gas mixture. The first oxygen was certainly produced by the photodissociation of water vapour exposed to ultra-violet radiation in the waveband 1500-2100 angstroms. Hydrogen would also have been produced, but this would escape comparatively quickly from the Earth's atmosphere. At first the oxygen would also have been rapidly removed as it reacted with the materials of the Earth's crust, but as these became fully oxidised, the level of free oxygen in the atmosphere would start to rise. Dr Harold Urey has recently pointed out, however, that the oxygen level in the atmosphere could not have risen to anything like today's 20 per cent as a result of photo-dissociation. The reason for this is that as free oxygen accumulates it tends to shield the Earth from the very ultra-violet radiation that is required to bring about the photo-dissociation. Thus the process is a self-regulating one, and Urey has calculated that the level at which the atmospheric oxygen becomes limiting does not exceed about 0.1 per cent of the present concentration. Ozone is also a particularly good absorber of ultra-violet and must have helped to shield the Earth from the lethal short wavelengths. As a confirmation of Urey's theory, results of the Mariner probes' analysis of the Martian atmosphere are in accordance with the presence of a self-regulating system such as suggested.

As the ultra-violet radiation was cut out by the accumulation of oxygen and ozone in the atmosphere, the synthesis of complex organic from simpler organic and inorganic substances would slow down and stop. But the polymerisation of self-replicating units in their coacervate droplets would be encouraged, and the organic units could grow much bigger, and would survive on or near the surface of the primitive seas.

The building up of the oxygen concentration in the atmosphere must then slowly have occurred as a result of the activity of autotrophic organisms over the subsequent millions of years.

Evolution becomes Irreversible

The cutting out of the ultra-violet bombardment of the Earth's surface was the point at which the evolution of life on this planet became irreversible. It meant that there was no longer any possibility of the synthesis of the amino-acids, sugars, polypeptides, etc., from which the first living organisms arose. It also meant that there was a strictly limited supply of usable organic raw materials that our ancestral coacervate drops could incorporate into themselves for growth and reproduction. Thus natural selection would have started to play an important part. The less stable and successful coacervate drops would have broken down again into their components, and these components would have been absorbed by the more vigorous drops and used for growth. Those drops capable of replication would have steadily become more numerous, and those without the power of reproduction would have gradually dwindled in numbers, as a variety of physical and chemical accidents befell them. Among the successfully replicating simple drops there would have been selective pressure in favour of those that had the power of absorbing and storing energy preferentially, and in particular those having the ability to synthesise complex and usable organic substances from simpler units, thus continuing the process performed previously by the ultra-violet radiation.

Clues from the Physiology of Present-day Organisms

The first successful autotrophes probably used solar radiation as their energy source for metabolism and synthesis. If they had had to rely on stored chemical energy it is difficult to see how evolution could have proceeded at the rate it did. This photochemical energy is nowadays made available to autotrophic organisms through pigments, which are iron, magnesium or copper atoms combined with porphyrins. They all absorb certain wavelengths of visible light, and accelerate hydrogen transfers, and oxidation reactions.

Taking synthesis first, many photosynthetic bacteria can build up carbohydrates from simple compounds using light energy. The following equations summarise some of the reactions carried out by these simple autotrophs:

$$CO_2 + H_2S \xrightarrow{\text{light}} (CH_2O) + H_2O + S$$

$$2CO_2 + H_2S + 2H_2O \xrightarrow{\text{light}} 2(CH_2O) + H_2SO_4$$

$$CO_2 + 2H_2 \xrightarrow{\text{light}} (CH_2O) + H_2O$$

$$CO_2 + 2H_2A \xrightarrow{\text{light}} (CH_2O) + H_2O + 2A$$

(Where A is a simple organic acid radical.)

Chemosynthetic bacteria, which obtain their energy from the oxidation of inorganic compounds, are also found today, but it is unlikely that they are primitive, both for the reasons mentioned above, and also because there can have been so little oxygen present at the time when the autotrophs were evolving.

The other important energy requirement is for metabolism, and as we have seen earlier, this is universally made available from stored energy in ATP. The conversion of radiant energy to chemical energy as ATP was therefore a vital step which must have evolved in the early autotrophs. If we consider the mechanisms by which present-day autotrophs produce their ATP, we can see the importance of the pigments once more. Calvin, who has recently elucidated most of the details of the photosynthetic process, has himself suggested that direct cyclic phosphorylation was probably the first step in the evolution of the photosynthetic mechanism. The dark and light reactions almost certainly evolved separately and then interlocked fortuitously. Once photosynthesis had arisen, involving as it does the splitting of water, the oxygen level in the atmosphere would have continued to build up to its present-day proportions, and thus permitted the development of aerobically respiring heterotrophs. The distinct autotroph-heterotroph balance would then have arisen, probably after a series of oscillations.

Conclusion

We have thus seen that within the framework of chemical and physical laws, and within the restrictions of the environment, it is possible that life evolved approximately according to the scheme outlined. There are obviously many steps which are not, and may never be clear and certain, and many alternative routes by which particular ends may have been reached, but the basic course outlined is quite a possible one.

Further Reading

Bernal, J. D. (1951) *The Physical Basis of Life*. London.
Calvin, M. (1956) "Chemical Evolution and the Origin of Life". *American Scientist*, **44**.
Haldane, J. B. S. (1933) *Science and Human Life*. New York.
Miller, S. L. (1953) *Science*, **117**.
Miller, S. L. (1955) *J. Amer. Chem. Soc.* **77**.
Oparin, A. I. (1938) *The Origin of Life*. New York.
Oparin, A. I. (1964) *Chemical Origin of Life on Earth*. Illinois.
Pirie, N. W. (1952) "Vital Blarney". *New Biology*, **12**.
Pirie, N. W. (1953) "The Origin of Life". *Discovery*, **114**.
Pringle, J. W. S. (1953) "The Origin of Life". *Symp. Soc. Exp. Biol.* **7**.
Urey, H. C. (1953) "The Concentration of Certain Elements at the Earth's Surface". *Proc. Roy. Soc. A.* **219**.
Wald, G. (1954) "The Origin of Life". *Scientific American*, **191**.
Bernal, Haldane, Pirie and Pringle all contribute articles on "The Origin of Life". *New Biology*, **16**. (1954.)

2
protistan radiation

It is now a well-established fact, easily confirmed by studying the fossil record, that during the course of evolution the diversity of living organisation gradually became greater and greater. It is a corollary of this fact that by extrapolating backwards we arrive at a point in time where there was no diversity. This is the point at which life originated, and it raises the problems that we discussed in the last chapter. Having once produced, theoretically at least, a coacervate drop capable of growth and replication, what was the next step? It is, of course, impossible to go back this far in the fossil record, as not only were the organisms we are concerned with microscopic, but they were also soft-bodied, and therefore only likely to be preserved in the finest grained sedimentary rock—if we are lucky. Also there are no unmetamorphosed sedimentary rocks still in existence more than 1,000 million years old, and we are concerned with creatures that existed at least another 1,000 million years before that. Therefore the only way in which we can guess at what these early forms of life were like is again by extracting the "Lowest Common Denominator" from organisms existing today, and obtaining as much guidance as possible from the organisms that today we consider to be the most primitive.

The least complex, and generally smallest forms of life today are found among the viruses, bacteria and protistans. Despite the definite terminology, these three types of organisation do merge into one another, and we are faced with the problem of deciding whether the similarities between viruses and bacteria and the similarities between bacteria and protistans represent convergent or divergent evolution. Viruses have the most clearly defined mode of life; they are all parasites; but this is not necessarily a valid reason for saying that they are not primitive in the evolutionary sense. The explanation of this is that most viruses operate by invading other living cells, and utilising their nucleic acids and amino-acids to synthesise more virus particles. This is exactly how our coacervate drops lived, though, except that they obtained their nucleic and amino-acids from the organic soup round them. It is therefore quite probable that when the supply of necessary organic substances in solution started to run out, some of those that were not capable of the critical syntheses continued to exist by parasitising the otherwise more successful organisms. It is thus possible that viruses are primitive, and have a common ancestry with all other living things, but that they are not ancestral themselves to anything but viruses and possibly a few organisms now classified as bacteria.

There is great danger in assuming that living organisms which we now consider to be low on the evolutionary scale today resemble the highest forms of life that existed, say, 1,500 million years ago. It is important to remember that the amoeba has been evolving as long as we have, and although it may bear far more resemblance to our common ancestor than you or I, the amoeba has been subjected to selective pressures for the same length of time as any other species of organism alive today.

Nevertheless, by considering the variety of protistans existing today, we may get some indication of the directions in which the earliest bacteria-protistans evolved.

Did one coacervate droplet give rise to every subsequent living organism on this planet, or did several equally successful droplets coexist originally and evolve into different branches of the living world? It is impossible to answer this question, but we can say that if there were many or several drops originally, then their physiological processes were identical as far as replication and energy provision are concerned, because today one cannot detect a polygenic origin of life.

Are autotrophs more primitive than heterotrophs? Again we cannot give a definite answer, as it is but a short step from absorbing complex organic substances in solution to absorbing the same constituents from inside an advanced coacervate drop or living organism. So it is quite possible that the autotrophic, holozoic, parasitic and saprophytic modes of nutrition have been coexistent from almost the beginning. It is also interesting to note that the normal definitions of these modes of nutrition are very difficult to apply at this early stage. Autotrophic organisms must have at least been dominant for some considerable time however, as, if the heterotrophs had gained supremacy at any stage, life would not have evolved any further once the initial supply of abiogenic organic materials had run out.

Let us now consider the range of protistan types found today, and endeavour to find any clues that may exist as to the types of radiation that occurred in the most primitive forms of life.

Conventional classifications of the Protista recognise four main classes:

1. Mastigophora or Flagellata e.g. *Euglena*
2. Sarcodina or Rhizopoda e.g. *Amoeba*
3. Ciliata e.g. *Paramecium*
4. Sporozoa e.g. *Monocystis*

This is a convenient but necessarily artificial classification. Thus the first three classes are distinguished by their mode of locomotion, and the fourth class, the Sporozoa, is to a large extent a taxonomic dustbin into which a large variety of parasitic but otherwise totally dissimilar forms have been thrown.

Which type of protistan is the most primitive? Despite the great resemblance of *Amoeba* and its relatives to our supposed ancestral coacervate drop, there is quite a strong body of evidence pointing to the fact that the Flagellata is the basal protistan group. The main points of evidence are as follows:

(a) Only among the flagellates is autotrophic nutrition found.
(b) The flagellum is a universally occurring organelle of propulsion, especially in the motile gametes and zoospores of higher organisms.
(c) "Young" forms of the Sporozoa and Sarcodina are often flagellate.
(d) "Young" forms of the Flagellata are never amoeboid, but adult forms occasionally produce pseudopodia. Their sporadic occurrence in various flagellate groups suggests that the amoeboid forms may have arisen several times from the basic stock.

A particularly interesting case is *Naegleria gruberi*, a protistan which changes readily from the amoeboid to the flagellate phase and vice versa according to changes in its internal pH. Corliss has suggested grouping all the Sarcodina and Flagellata together in one class, the Mastigamoebaea, but this is probably less informative evolutionarily than his other step of elevating a group of the Sporozoa to class status. This class he calls the Cnidosporidia.

These organisms are multinucleate parasites which at one stage in their life history possess pole capsules, organelles bearing a marked resemblance to coelenterate nematocysts. This is a very intriguing fact, as nematocysts are found nowhere else outside the coelenterate phylum except in certain molluscs and platyhelminths, which are now known to obtain their complement of stinging cells from the coelenterates on which they feed. There is therefore a strong suggestion of relationship between the coelenterates and cnidosporidians. That the latter evolved into the former is entirely untenable, because of the specialised parasitic modes of life of the cnidosporidians, but it is becoming more generally accepted now that these "protistans" may in fact be greatly reduced parasitic medusoid forms, in which case they are really coelenterates.

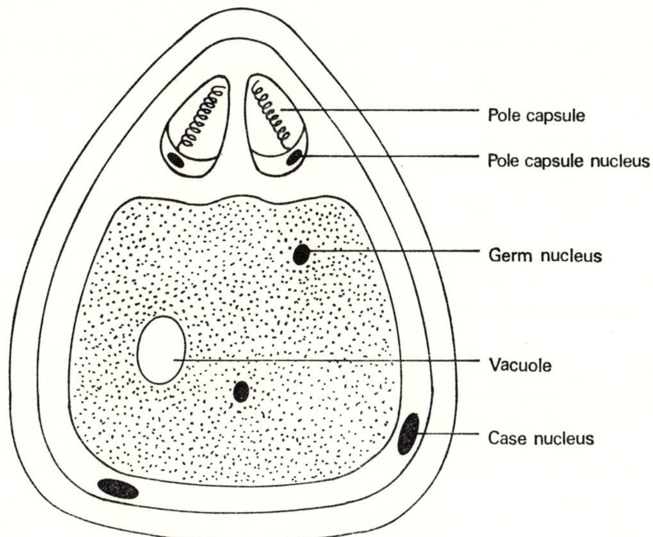

Fig 2.1. A cnidosporidian spore

This example is mentioned not only for its intrinsic interest, but also to emphasise the fact that even among the protistans, things are not as simple as they may seem.

The Origins of Multicellular Organisation

We have until now carefully avoided referring to the Protista as either unicellular or acellular. This is, of course, because they can be either, according to one's definition of the word 'cell". If we use the word in its original sense as being one of the small units into which most higher organisms are divided then the protistans are acellular. If, however, we consider the cell as being one of the units that make up an organism, then the protistans are one-celled.

Is either of these definitions more biologically correct than the other? As we shall see later, evolutionary considerations may help here, but first let us consider the problem of the origin of the many-celled organisms.

Multicellular organisation means decentralisation and specialisation. Thus in contrast to the Protista, where even in the multinucleate forms no nucleus ever has sole command over a specialised part of the cytoplasm; in the metazoa and metaphyta the many nuclei each have charge of a portion of cytoplasm which has some specialised function.

In our present society, it is easy to doubt the efficiency of specialisation, but the pressures which produced it in society are the same as the pressures which brought about its inevitability in the evolving organism, namely those resulting from increase in size.

This latter phenomenon is a tendency that has expressed itself throughout the course of evolution. There have been numerous exceptions, but in most of the major groups the tendency to become larger is very marked until an optimum size is produced. We must conclude as a result of this that a relatively large size is of some selective advantage in inter- and intra-specific competition, and within limits, this is a very reasonable conclusion. Obvious benefits resulting from an increased size are physical superiority over smaller organisms, and a greater degree of independence of the environment. Thus the larger the organism, the larger has to be the environmental fluctuation (in temperature, humidity, osmotic pressure, etc.) that will catastrophically affect the organism. Together with this greater independence may go an improved ability to manipulate the environment, but this is not necessarily so. Either of these benefits, however, will enable the organism to control its "milieu interieur" to a greater extent and therefore improve its chances of "la vie libre".

It was noted earlier that specialisation, and multicellular organisation were a direct result of increase in size. Why should this be so? The answer is concerned with the physical limitations imposed by increase in size. Let us consider the case of a hypothetical spherical protistan that increases in size in such a way that its shape remains constant but its radius doubles. By using simple mathematical formulae we can calculate that as a result of this growth its surface area will have quadrupled and its volume will have increased eightfold. The practical consequences of this are that:

1. The diffusion path from the surface to the inner cytoplasm has doubled in length.
2. The volume of respiring cytoplasm and nucleoplasm has increased eight times, whereas the surface area available for respiratory exchange has only increased fourfold.
3. The mass of the organism has increased eightfold.

These facts would obviously effectively prevent such a size increase passing a certain critical point, at which one or more of the above factors would become limiting. At this point specialisation becomes necessary, and an increased respiratory surface, a circulatory system and some form of skeletal system are vital developments if any further evolution in size is to occur. These further developments involve differentiation, and although a certain amount of specialisation can and does occur within the protistan body, above a certain degree of complexity, cellular organisation becomes inevitable.

Phylogenetic Origins

How then have the Metazoa and Metaphyta arisen? Three main hypotheses have been put forward to explain the origin of the multicellular grade of organisation.

1. The first theory, which has many variants, supposes that the metazoa arose by the proliferation of a single protistan to form a colony, which then underwent a process of individuation to produce a single organism. There is a marked tendency for certain colonies to individuate, as is well shown by the siphonophore coelenterates (e.g. *Physalia*) and some colonial algae (e.g. *Volvox*).

2. The second theory derives the Metazoa from the Protista by means of internal divisions occurring in multinucleate ciliates.

3. The third theory seeks to derive both the multicellular organisms and the Protista from a common multinucleate ancestor.

Of these three theories, the first was the most popular for a considerable time in the form in which it was propounded by Haeckel. This was based on his so-called "Biogenetic Law", according to which; "ontogeny is a short recapitulation of phylogeny". In other words, each organism in the course of its embryonic development passes through a series of stages, each of which represents the adult form of a stage in its evolutionary development. Thus the visceral clefts present at one stage in embryonic mammals represent, according to this theory, the gill slits of the adult ancestral fish. Haeckel further suggested that the fertilised ovum represented the protistan stage in evolution; the blastula represented the Volvox-like colony; and that the gastrula represented the coelenterate stage, with two germ layers and an archenteron with one opening. Superficially the story appears to be so neat as to be manifestly true, but it had not been long published before attacks were made on it from many angles, and quite clearly in this century the majority of biologists had come to accept the view that while the embryonic stages of living animals may represent the embryonic form of their evolutionary ancestors, there is no question of them resembling the adult form of the ancestor. The part of the theory that explained the origin of the Metazoa, however, was retained and remained virtually unchallenged until 1944. Thus it was believed that the ancestral protistan divided to form a hollow ball of cells resembling the blastula stage of many animals, which was accordingly called the Blastaea. This, it was considered, then underwent a process of invagination to form a double layered structure, correspondingly called the Gastraea. It followed from this therefore, that the Coelenterata represent the most primitive metazoan phylum, being diploblastic and largely radially symmetrical.

In 1944, however, Hadzi produced an alternative theory of metazoan origin, which is the second one outlined above. He believed the internal division and cellularisation of a multinucleate protistan to be a much more likely origin of Metazoa than the route via blastaea and gastraea. No loss of integration in the organism would be involved this way, and the necessity for individuation of the colony-blastaea stage would be dispensed with. Furthermore, he pointed out, the Coelenterata are not strictly diploblastic, as cells are always found in the mesogloea to a greater or lesser extent. Hadzi then proposed that the turbellarian Platyhelminthes were in fact the most primitive metazoan group, and had arisen by the internal division of multinucleate protistans. Bilateral symmetry, he claimed, was already present in most of the more

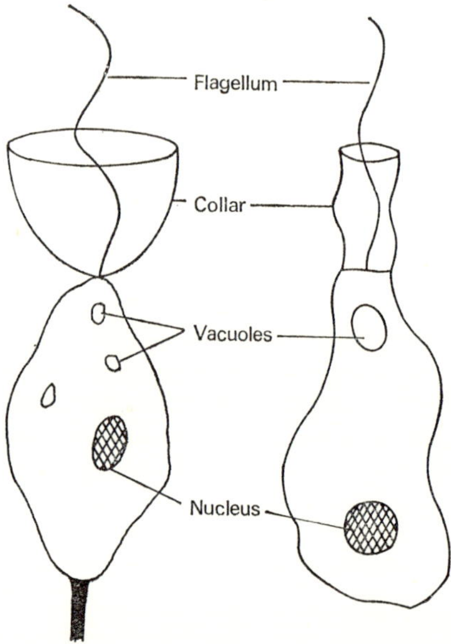

Fig 2.2. A choanoflagellate. A sponge choanocyte

advanced protistans, and the Coelenterates probably arose from the rhabdocoel turbellarians after some of them had adopted a sessile habit. The Anthozoa on this basis would be the most primitive coelenterate class as they are bilaterally symmetrical, and have no medusoid stage. Hadzi's theory is nowadays quite widely accepted, and many additional pieces of evidence have been produced which confirm that the turbellarians are likely to be the most primitive metazoans; and that the coelenterates are anything but as simple as their structure might lead one to suppose they were.

The third main theory mentioned above, that both Protista and Metazoa are descended from a common ancestor, has not much practical evidence to support it, and in one form differs very little from Hadzi's theory of origins from a multinucleate protistan.

Apart from the plant kingdom, whose origins and evolution shall be considered separately in later chapters, there remains one group of multicellular organisms which almost certainly had an origin different to that of the rest of the Metazoa. This group contains the sponges, and is known as the Parazoa. Although many celled, these organisms differ so much from the rest of the Metazoa, that they are often given an equal status with the Protozoa and the Metazoa, that of a sub-kingdom. The reasons for this are many, the main ones being that there is no co-ordinating system of any kind linking the individual cells, and that the cells themselves are of a peculiar type known as choanocytes. The cells give us a clue to the origin of the group, for these choanocytes are almost indistinguishable from the members of a family of flagellates, called the choanoflagellates, and it seems highly probable that the sponges arose from a colonial form of these flagellates.

Thus multicellular animal organisation probably arose twice from the Protista; once by proliferation and individuation to give the Parazoa; and once by internal division of a multinucleate ciliate to give the Turbellaria and thence the rest of the Metazoa.

Finally, to return to the definition of the Protista. If Haeckel was right, we should perhaps use the term "unicellular", whereas if Hadzi's theories are correct, "acellular" may be a more justifiable epithet.

Further Reading

Borradaile, Eastham, Potts and Saunders. (1961) *The Invertebrata.* Cambridge University Press.

Clark, R. B. (1964) *Dynamics in Metazoan Evolution.* Oxford University Press.

Corliss, J. O. (1956) "On the Evolution and Systematics of Ciliated Protozoa". *Syst. Zool.* 5.

De Beer, G. R. (1954) "The Evolution of the Metazoa". *Evolution as a Process.* London.

Hadzi, J. (1953) "An Attempt to reconstruct the System of Animal Classification". *Syst. Zool.* 2.

3
the earliest metazoa

In the last chapter, we deduced theoretically that the most primitive animal phylum is the Protista. We also saw how certain metazoan stocks probably arose from these primitive forms. In contrast to these hypotheses, one may ask, what does the fossil record indicate about these early stages? The answer is that at the moment the picture presented by these early fossils is necessarily limited. The earliest known fossils are from Pre-Cambrian rocks situated in various parts of the world, and the animal forms represented by these include coelenterates, brachiopods, annelids and echinoderms. So although the remains are sparse, a considerable degree of radiation had occurred as long ago as 700 million years. The reasons for the paucity of these early remains are several. One major factor is that Pre-Cambrian rocks themselves are rarely accessible, and of those that are, a large proportion are igneous or metamorphic, and therefore likely to be completely unfossiliferous. The unaltered sedimentary rocks of this era, however, are also almost devoid of fossils, because until Cambrian times no animals had developed hard internal or external skeletons to any great degree, and therefore the organisms at death were unlikely to be preserved in any but the very finest grained sedimentary rocks, and then only as two-dimensional impressions.

Thus it is true to say that the first radiation of the major phyla is as unrecorded as was the origin of life itself, and therefore once again one has to deduce the course of events from phylogenetic, embryological and physiological studies. Before attempting to do so, however, it is worthwhile considering the phyla that did appear during the Cambrian, for it is then possible reasonably to deduce which organisms were most successful at that time.

The Cambrian Fauna

Somewhere between 500 and 700 million years ago, the era known as the Cambrian began, and during the course of the next 100 million years or so before the close of the era, seven animal phyla appeared. This does not mean that they evolved into differentiated phyla during this time. This process had almost certainly occurred during the Pre-Cambrian period, but they appear in relative abundance in the fossil record. Their order of appearance is not particularly significant, nor is their relative frequency, for as we have seen, some organisms inhabit regions far more likely to preserve an abundance of fossil

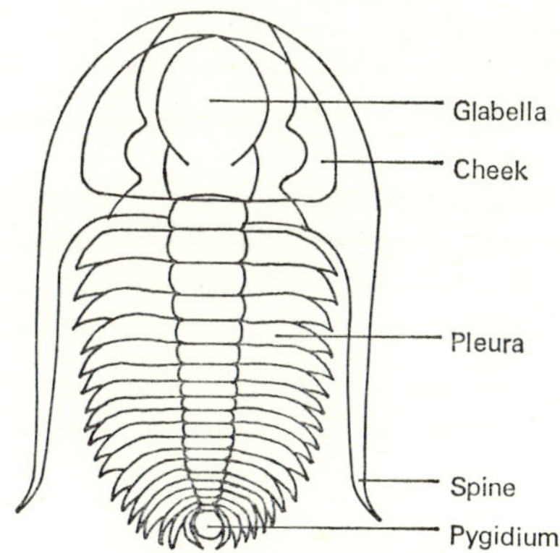

Fig 3.1. Paradoxides

material, and the apparent frequency may reflect no more than a well-marked exoskeleton, which will also predispose its possessor to preservation. It is significant, however, that no animals above the invertebrate grade of organisation had appeared, and it is possible to follow the evolution of the vertebrates through systems subsequent to the Cambrian.

The most characteristic fossils of the Cambrian strata are the early trilobites. These are the first arthropods to appear in the fossil record, and although totally extinct now, they were undoubtedly the dominant form of life in the Cambrian seas. Over 700 genera are known, ranging in size from less than half a centimetre to over 50 centimetres long. Their basic structure is illustrated above, as exemplified by *Paradoxides*, one of the larger genera.

Other contemporary softer bodied arthropods are also known from the Mid-Cambrian Burgess shales of British Columbia.

The second phylum with major representation in the Cambrian period is the Brachiopoda. The animals in this phylum are bivalved "shell-fish" (not molluscs) which live attached to the sea-bed and feed by filtering particles of nutriment from the water. They are unique in that one genus, *Lingula*, has persisted apparently unchanged from late Cambrian times to the present day.

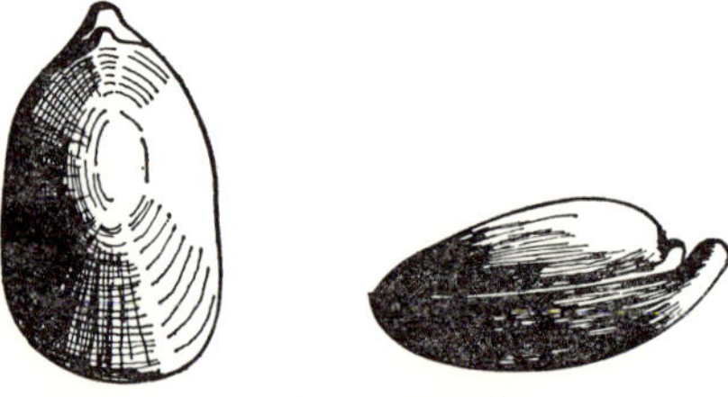

Fig 3.2. Lingula

Other phyla represented in the Cambrian seas were:

1. Porifera. The sponges.
2. Coelenterata. Mainly as jellyfish.
3. Mollusca. Mainly gastropods.
4. Annelida. Sometimes preserved entire, but more frequently represented by tracks, burrows and chitinous jaws.
5. Echinodermata. Mainly representatives of extinct classes.

Finally it must be assumed that algae were also present universally and abundantly as primary sources of food. They are known from the Burgess shales, but elsewhere cannot have been preserved due to their delicate structure.

The Times Change

The Cambrian period lasted for about 100 million years and at the end of this time changing conditions produced such major alterations in the climate, marine topography and subsequently the fauna that geologically and palaeontologically a new name is justified for the following period. It is known as the Ordovician.

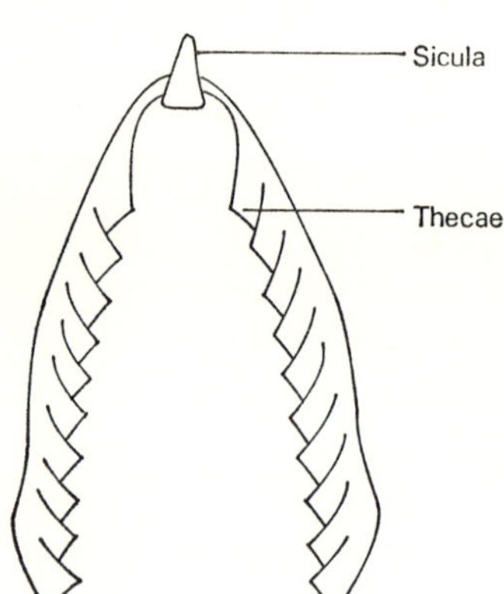

Fig 3.3. Didymograptus

Lasting about 60 million years, the Ordovician was a time of warm shallow seas, with marked geosynclines in some areas giving troughs of deeper water. Volcanic activity was marked. As a consequence of this varied state of affairs, Ordovician strata today differ widely in composition and in faunal content. The shallow water shelly deposits contain later forms of trilobites and brachiopods, with molluscs and echinoderms still present. In addition to these, however, corals make their first appearance, and three new phyla are found; the Protozoa, the Bryozoa and the graptolites. This latter group is in fact much more characteristic of the deeper water shale deposits, as graptolites (now extinct) apparently led a planktonic existence.

Graptolites were originally considered to be variously plants, coelenterates or bryozoans, but now they are generally accepted to be closely related to the early chordates.

The Ordovician volcanic strata, as one might expect, are completely unfossiliferous.

Origin of the Palaeozoic Invertebrate Phyla

As we have already seen, the divergence of the first invertebrate phyla found in the fossil record occurred in Pre-Cambrian times, and we cannot hope to find a record of this evolutionary radiation. Also it is a fact that in the 500 million years which have elapsed since the appearance of these phyla, their basic organisation has remained completely unchanged. *Lingula* is an example of a genus which has persisted throughout this period of time. It is thus impossible to deduce by backward extrapolation the probable points of divergence of these phyla. One is presented with a *fait accompli* in the Cambrian, and other criteria must be used to arrive at possible theories of common ancestry.

The Superphyla

When listing defining characteristics of invertebrate phyla, certain fundamental diagnostic features must be included. These are:

1. The number of cell layers. (Diploblastic or triploblastic.)
2. Whether a coelom is present.
3. Whether a haemocoele is present.
4. Whether segmented.
5. Whether a larval stage occurs, and if so, of what type.

Other characteristics tend to be diagnostic of one or only a small group of phyla (e.g. the presence of nematoblasts in coelenterates) and may be the feature or features which have ensured the evolutionary success and distinction of a phylum.

The features listed above are all reasonably non-adaptive and may be of value in deciding true affinities.

Number of Cell Layers

As we saw in the last chapter, if Hadzi's theory as to the origin of the Metazoa is correct, the triploblastic grade of organisation is the most primitive. There is nothing specially significant or magic about these three layers, however, as has sometimes been claimed, for if cellularisation of any animal with a gut occurs, three layers of cells are bound to be produced: one layer comprising those cells in contact with the outside world; one layer composed of those cells lining the gut; and the third layer (mesoderm) consisting of the remaining cells between the ectoderm and endoderm. The bounding layers then become specialised for their functions of protection, and digestion and absorption respectively, and the intermediate cells differentiate into other organs, glands and tissues as necessary.

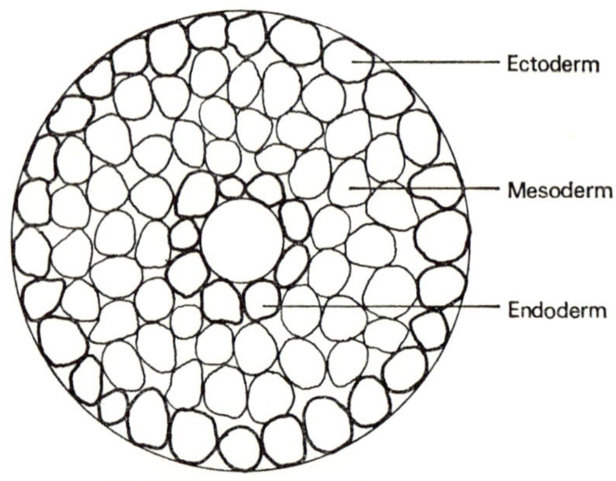

Fig 3.4. The formation of cell layers in a metazoan

This means that the diploblastic type of organisation, as found only in the coelenterates, is a secondary simplification, and it means that the most purely diploblastic classes are the most specialised.

Presence of Coelom

The development of some kind of body cavity was a feature which contributed greatly to the success of its possessors. Thus of the major triploblastic phyla, only the platyhelminths, nematodes and rotifers are acoelomate, although in the arthropods and molluscs the coelom is reduced and its functions taken over largely by the haemocoele. The chief advantages conferred by the possession of a coelom are:

1. Easier movement because of the hydrostatic skeleton properties of a fluid filled cavity.
2. Separation of peristaltic movements of the gut from movements of the body wall.
3. Its function as a simple circulatory system for transferring food substances from the gut to the rest of the body, and excretory products to the outside via coelomoducts.

Described in its simplest terms, a coelom is a fluid filled cavity surrounded by mesoderm. It may, however, develop embryologically in more than one way. Thus in *Amphioxus*, following gastrulation, the mesoderm separates from the endoderm as a series of pouches, each of which encloses a portion of the archenteric cavity which becomes the coelom. In other chordates, on the other hand, the coelom arises *de novo* from a split in the embryonic mesoderm. These two conditions are described as enterocoelic and schizocoelic respectively.

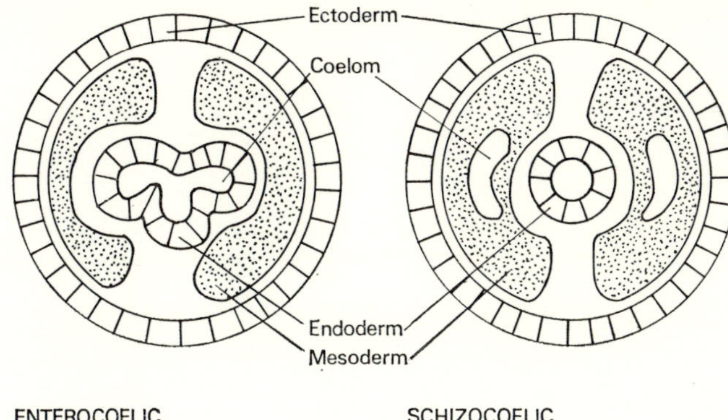

Fig 3.5. Two different methods of formation of the coelom

Presence of a Haemocoele

A haemocoele is found in some coelomates, where the functions of the coelom are taken over by a greatly enlarged blastocoel filled with blood. A well-developed haemocoele with full hydrostatic skeleton functions is only found in the arthropods and molluscs.

Segmentation

When as a result of developmental processes an organism can be clearly seen at some stage to consist of a number of more or less identical serial components, it is said to be segmented. Except for the very simplest examples there is always differentiation between different segments, but it is possible to derive these varied components from a basic embryonic segment with serially repeated organs. Often this segmentation is apparent externally as in the annelids, but sometimes it is necessary to investigate internal anatomy, or early embryonic stages in order to observe the segmented structure. This is true of the higher chordates.

Segmentation must not be confused with strobilisation, which is the serial repetition of proglottids as seen in the tapeworms. Here there is no coordination between "segments", and the proglottids at the posterior end are constantly being shed when mature, whilst new proglottids are being produced from the scolex. This process is rather analagous to budding.

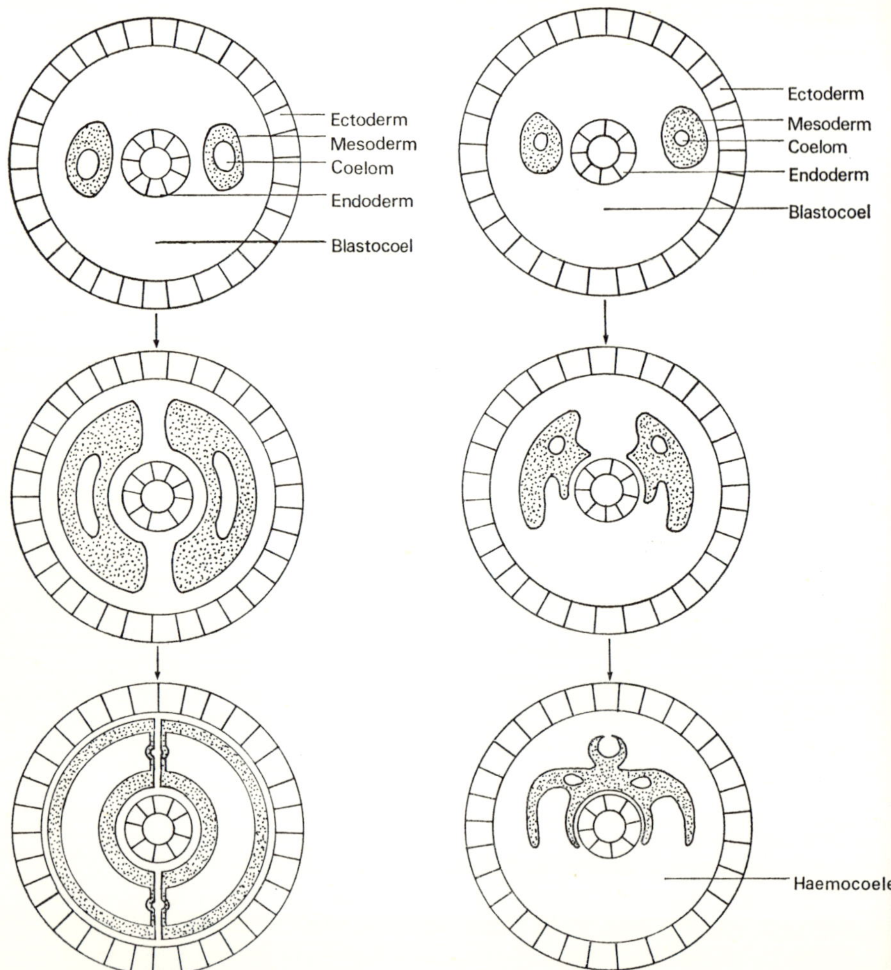

Fig 3.6. Development of the coelom as seen in annelids

Fig 3.7. Development of haemocoele as seen in molluscs and arthropods

Larval Stages

In many groups within the animal kingdom it is found that the fertilised ovum, instead of developing directly into the adult form of its species, first produces a form which may be distinctly different from the adult in many respects. This stage is known as a larva, and at some point in its life, there is a more or less rapid change from this larval stage to the adult form. This process of change is a metamorphosis.

The significance of a larval stage is that it differs radically from the adult in one or more aspects of its ecology, and therefore it also differs anatomically, physiologically and behaviourally.

A common point of difference is that the larval stage is motile, whereas the adult is sessile. This ensures distribution of the species. Other common differences are that the larva may be aquatic and the adult terrestrial; or the larva may feed whereas the adult is incapable of feeding, being purely reproductive.

The evolutionary significance of larval stages is enormous, however, partly because great advances may occur through precocious development of the larval stage and suppression of the adult as we shall see

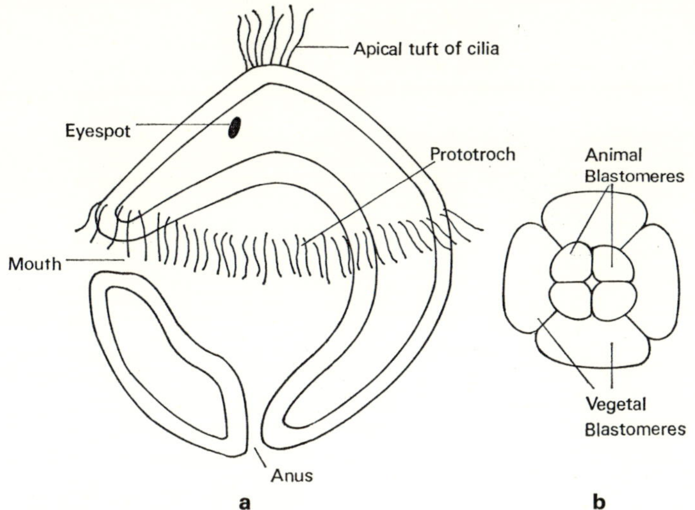

Fig 3.8. a—Trochophore
b—Spiral cleavage as seen looking down on the animal pole

later; but also because larval stages tend to remain relatively unaltered when large structural modifications occur in the adult. Thus it is possible to recognise affinities between two apparently dissimilar groups of animals comparing their larval stages.

All metazoans, apart from those already discussed (the Porifera, Coelenterata and Platyhelminthes), can be recognised on larval and other criteria as belonging to one of two major groups or "superphyla" as they are known; the annelid superphylum and the echinoderm superphylum.

The Annelid Superphylum

The animals in this group are similar in the following basic respects:

1. Where a larval stage exists, it is of the trochophore (or trochosphere) type. This is a more or less spherical larva, which shows bilateral

symmetry as a result of the position of its mouth and anus, and which has an equatorial band of cilia which it uses for swimming, called the prototroch.

2. During the process of cleavage, the first two divisions of the fertilised egg are always vertical and mutually at right angles. The third division is then equatorial. At the end of these first three divisions, it is a characteristic of all members of this superphylum that the four upper (animal) blastomeres lie not directly on top of the four lower (vegetal) blastomeres but over the junctions between these vegetal cells. This course of events is known as spiral cleavage, and

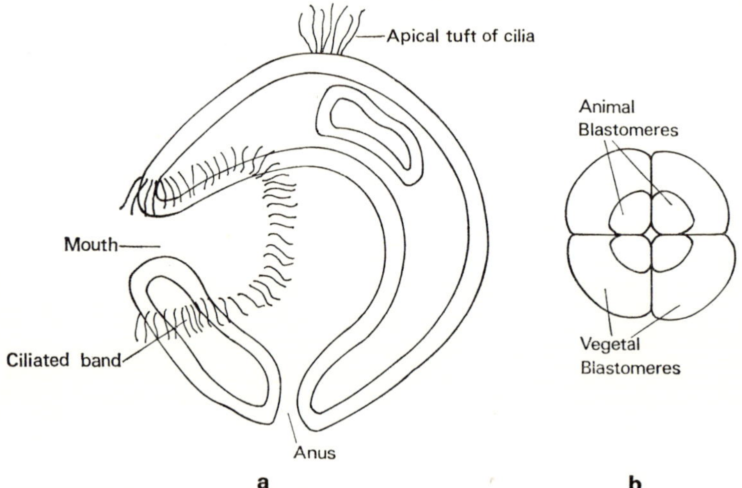

Fig 3.9. a—Pluteus larva
b—Radial cleavage as seen looking down on the animal pole

is not due to the animal blastomeres shifting round 45°, but to the oblique orientation of the mitotic spindle with respect to the animal-vegetal axis during the third division.

3. The difference between a schizocoelic and an enterocoelic coelom has already been explained, and it is the former which is invariably found in members of the annelid superphylum. The coelom in these, therefore, never has contact with the cavity of the enteron.

The Echinoderm Superphylum

The main characteristics of this group of animals are as follows:

1. The larval stage, when present, is of the pluteus type. This is very similar to the trochophore, but the ciliated band which is used for swimming and feeding forms a ring round the mouth instead of an equatorial band.

2. Cleavage is radial. This means that following the third division of the egg, the animal blastomeres lie directly on top of the vegetal blastomeres.
3. The coelom is enterocoelic, although the process of development may be greatly modified.

Bearing in mind the points mentioned as being diagnostic of these two superphyla, we can consider the major phyla in turn and attempt to place them in one or the other group.

Platyhelminthes

The flatworms are bound to be rather difficult to place, as they are primitive in several ways. Thus they are acoelomate, and therefore it is obviously impossible to classify them according to the mode of formation of the coelom. They also have only one opening into the gut, and hence their larval stage (Muller's Larva) differs from the trochophore in having no anus. It does, however, have an equatorially situated ciliated band, and the cleavage is radial, suggesting that this phylum has affinities with the annelid superphylum. The Platyhelminthes probably diverged from this main stock before the development of the coelom or the anus, and it is of interest to note that they did not evolve a coelom subsequently.

Annelida

The annelids comprise the phylum on which the annelid superphylum is based. No more, therefore, need be said here, as they are typical in every way.

Arthropoda

The arthropods have become very specialised and none have an aquatic larval stage resembling the trochophore. The coelom is also very much reduced, and a haemocoele is present. It is certain from other evidence, however, that the arthropods and annelids are evolutionarily closely related, and therefore the arthropods must be included in the annelid superphylum.

Mollusca

The members of this phylum have a larval stage which is an indisputable trochophore. Cleavage is radial and the (reduced) coelom is formed schizocoelically. Thus although the adult molluscs show such a great diversity of form, all very unlike other members of the annelid superphylum, the larval and developmental characteristics of this phylum leave no doubt as to its true affinities.

Echinodermata

This is the type phylum of the echinoderm superphylum, demonstrating all the main characteristics.

Chordata

There are no larval stages resembling a pluteus found anywhere in this phylum, but cleavage is spiral and an enterocoelom develops (although this may appear to be superficially schizocoelic amongst more advanced members of the phylum). There are also considerable further links between the chordates and the echinoderms, as we shall see later.

Other Invertebrate Phyla

The nemertine worms have a larva very similar to that of the Platyhelminthes, and should probably be placed in the annelid superphylum. It is interesting to note in these animals, however, that although the larva has no anus, the adult develops one.

Some other phyla, such as the Rotifera and the Nematoda are so greatly modified that it is impossible to assign them to either superphylum with any degree of accuracy. In some other cases, however, such as the Polyzoa and Chaetognatha, there are clues as to affinities which may be mutually contradictory.

Conclusions

Having seen that the major phyla can be grouped into two superphyla, what evolutionary conclusions can we now draw? The main one is probably that at a very early stage in the evolution of animals, two main stocks diverged, each with a characteristic aquatic larva, and each with a distinctive method of development. During the subsequent radiation of the adult stages of these two stocks, phyletic differences evolved, and there was also considerable variety at an infra-phyletic level. In some cases the developmental stages were also modified or shortened, and in these cases it is impossible to deduce affinities. But more often the early stages remained remarkably constant, whilst vast morphological and physiological changes in the adults produced forms ranging from flatworms to butterflies on the one line, and from starfish to Man on the other.

It is particularly interesting to note that the coelom, as a body cavity formed within the mesoderm, must have arisen independently in the two main evolutionary lines, thus emphasising the strong selective advantage endowed by the possession of such a cavity.

The concept of a superphylum does not, however, indicate evolutionary lines within these two main groups. Thus it is reasonable to assume that the acoelomate Platyhelminthes diverged from the annelid superphylum stock at a comparatively early stage, but we cannot tell at what stage the molluscs and annelids developed, nor what the ancestral forms were like. We shall consider the origin of the arthropods and the radiation within the main invertebrate phyla in the next chapter, but for most groups the phyletic origins and relationships are extremely obscure.

Further Reading

Borradaile, Eastham, Potts and Saunders. (1961) *The Invertebrata*. Cambridge University Press.

Carter, G. S. *A General Zoology of the Invertebrates*. London.

Clark, R. B. (1964) *Dynamics in Metazoan Evolution*. Oxford University Press.

Schrock, R. and Twenhofel, W. (1953) *Principles of Invertebrate Palaeontology*. New York.

Wells, A. K. and Kirkaldy, J. F. (1968) *Outline of Historical Geology*. Allen and Unwin.

4

evolution of the annelida, mollusca and arthropoda

The fossil record provides a certain amount of information about evolutionary sequences and affinities, but it is of necessity very limited and possibly misleading, unless interpreted carefully. Embryology and physiology also provide what is probably more helpful information, and we have seen that the concept of the two superphyla is derived almost exclusively from these studies. The problem that continually besets the evolutionary taxonomist, however, is which structural or physiological features are fundamental, and which are adaptive.

In recent years, a number of researchers have turned their attention to what might be called an ecological approach to invertebrate affinities, with the result that they emphasise the important point that it is impossible to distinguish between "taxonomic" and "adaptive" features in many invertebrate groups. Manton in particular has done elaborate and exhaustive work on arthropods, especially myriapods, and she emphasises the artificiality of the dichotomy between the classificatory and functional significance of certain characters.

On a phyletic level, structural adaptations to a particular locomotory habit may more or less obscure the fact that two organisms occupying the same environment and having the same habit are in fact totally unrelated. This is convergence. Conversely, two closely related organisms occupying the same environment but having different habits may appear to be very different structurally from one another and therefore to be unrelated.

Examples of these two situations are numerous, but may well be exemplified by the many groups of animals that have adopted a long cylindrical "worm" shape as an adaptation to burrowing; and as an example of divergence, the adoption of a sedentary habit by the adult

Cirripedia makes it very difficult to recognise a barnacle as being a crustacean.

The Coelom Reconsidered

Even apparently unadaptive characters can often be shown to be adaptive to a particular habit when considered ecologically. Thus we have already seen that the coelom, however it is derived embryonically, confers certain advantages on its possessor. Could it, however, ever have conferred a specific advantage in a particular habit?

If we return to the point in the evolution of metazoans where a trend to an increase in size had reached the stage where the basic protistan locomotory organelles were no longer efficient, it is clear that the musculature of the body wall must have developed then. If these muscles are to allow reversible changes of body shape to occur, the only arrangement for them is in alternating layers of circular and longitudinal muscles. This will allow movement by means of peristaltic waves and also by various forms of looping, but the movement will only be able to take place on a surface such as the sea-bed. The circular muscles can still not play a very big part in locomotion, however, as the moment they become operative, the body shape tends to become rounded, with a consequent lessening of the area of the organism in contact with the ground, and therefore a lessening of locomotory efficiency.

The situation in the early seas at this point was that there was a large free-swimming fauna, and probably a large bottom-living fauna, but at this acoelomate stage of animal evolution, there were no organisms in existence with the mechanical capabilities of burrowing deeply into the loose deposits at the bottom of the seas. The advantage of being able to burrow would presumably at first have been mainly concerned with protection against the predatory benthic (bottom-living) fauna, but apart from this, the burrowing habit was exploiting an entirely new environment.

With the two muscle layers already in existence, all that was necessary for the first animals to become efficiently adapted to burrowing was the development of some form of hydrostatic skeleton, and it must have been here that the coelom first evolved and was selected for. We have seen already that coeloms can arise in two very different ways. In fact there are at least five distinct ways in which the structure known as a coelom can develop embryologically, and this fact has caused a lot of debate as to the true origin of the body cavity. Given the situation described above, however, it is entirely reasonable to postulate that any body cavity, no matter how it was derived, would be selected for, and it seems likely therefore, that the coelom may well be polyphyletic in origin.

The Annelida

Present-day annelids are mostly considered as belonging to one of the three main classes: the polychaetes, the oligochaetes, or the hirudineans. The main distinctive characteristic of this group of coelomate invertebrates is that they are obviously segmented, a feature which persists to a greater or lesser extent in all higher phyla. What are the possible advantages conferred by this form of metameric segmentation?

Metamerism

We have seen that some kind of hydrostatic skeleton is necessary for the adoption of the burrowing habit, given the existence of layers of circular and longitudinal muscles. In order to localise the effects of the action of these muscles, however, some subdivision of the coelom is required. This then permits the sequential changes of shape of parts of the body which are necessary for efficient and sustained burrowing as seen in the earthworm. Thus the evolution of septa must have been an important first stage in the development of metamerism. The segmentation of the body wall musculature, the nervous system, and the blood vascular system would have followed, as the "co-ordinated independence" of the body regions developed.

Evolution within the Annelida

The two contenders for the title of the most primitive class of annelids are the polychaetes and the oligochaetes, as the leeches are obviously very specialised for their mode of life. If we draw up a list of features that are generally agreed to be primitive, then it may be easier to decide between the polychaetes and the oligochaetes.

The earliest annelids were probably marine, with unmodified metamerism, and simple chaetae (which would have assisted considerably with burrowing). They were almost certainly unisexual, with external fertilisation, and a free-swimming trochophore larva. Also, their segmental organs would have been serially repeated with no condensation or concentration of reproductive organs, for example, into one region of the body.

Superficially it is very tempting on these criteria immediately to place the polychaetes in the more primitive position, as they fit the above supposed primitive characters almost exactly. The oligochaetes, on the other hand, are clearly very specialised according to these criteria, and therefore presumably more advanced than the polychaetes. If, however, we consider the fact that the oligochaetes are either fresh-water or terrestrial in habitat, and then take into account the modifications from the primitive form necessitated by this change in habitat, it becomes clear that all the "advanced" features can be considered as necessary modifications. Similarly, those "primitive" features of the polychaetes are mainly adaptations to a marine life.

Leaving aside all features which are clearly adaptations to mode of life and habitat, we are left with very few morphological and anatomical characters on which to base a judgement. One of these characters, the arrangement of the chaetae, has been used in support of arguments favouring each class as being the more primitive of the two, so at the moment, the question is very much open.

The Mollusca

The molluscs comprise one of the most fascinating groups of invertebrates. Quite apart from the interest of the individual species, the relationships between the classes and the origin of the phylum itself are very unclear. This is well demonstrated by the fact that the discovery of a single "living fossil" species in the 1950s has completely challenged existing theories about the origin of the phylum.

The range of form and habitat within the group is large and diverse, from the sessile limpets at one end to the fast swimming squids at the other. Such were the difficulties in attempting to derive all molluscs from a common ancestral stock, that a hypothetical "archimollusc" was invented, which was a lowest common denominator of molluscan features.

As far as phyletic relationships are concerned, a study of the embryological development of molluscs suggests that they are more closely related to annelids than to any other phylum. This is because the stages of development up to the formation of a trochophore larva are virtually identical in the two phyla. And yet almost all living molluscs have a haemocoelic body cavity and are unsegmented.

In 1957 a new species of living mollusc was described from the deep water off the Mexican coast. This was *Neopilina*, a species with a shell almost identical to that of *Pilina*, a species only known as a fossil, and not known to have occurred since the Devonian period. This living fossil is remarkable in many ways, but the most shattering anatomical feature about it is that its gills, gonads, excretory organs and shell muscles all occur in serially repeated pairs. Whether this repetition represents the remains of true metamerism or not is still a disputed question, but it opens up the possibility that molluscs are descended from a metamerically segmented ancestor, possibly descended in turn from a common ancestor with the annelids.

Evolution within the phylum well illustrates adaptive radiation. Thus each of the five main classes of molluscs has a very distinctive habit, and has become anatomically and physiologically modified to that habit, hence emphasising the differences between the classes. If we consider the three best known mollusc classes we can see this clearly.

The Gastropoda, forming the largest class of molluscs, move by crawling on a flattened ventral foot. They mostly have a dorsal shell into which they can withdraw for protection, and they are mostly grazing feeders, using a radula to rasp off their required food material. Two very interesting evolutionary trends are seen in the gastropods, one of which is more or less caused by the other.

The major trend is the tendency to torsion. This is a process by which during the larval stage of many gastropods, the visceral hump is swung through 180 degrees so that the mantle cavity, gills and anus which originally were posterior, become anterior facing. The advantages of this are not particularly clear, as some gastropods manage well without having undergone torsion, but there are two main theories as to its advantages. One of these, propounded by Garstang in the form of a delightful poem, is that torsion would have benefited the larval mollusc by providing an anterior space (the mantle cavity) into which the sensitive head could be withdrawn quickly in an emergency. This theory sounds very plausible but is unlikely to be true because the main enemies of larval gastropods are plankton feeding fish which would hardly be deterred from their diet by the retraction of the larval molluscs' heads. The second and more probable theory as to the advantages of torsion is that the advantage would have been to the adult. Thus the mantle cavity becomes a sense receptor for determining the conditions prevailing in the water in the region of the head of the animal.

The second main evolutionary tendency observed in gastropods is to the coiling of the visceral mass. This is necessitated to a certain extent by torsion and results in the familiar spiral structure seen in land gastropods.

The Bivalvia, a second molluscan class, differs markedly from the gastropods. Whereas the members of the latter group are free grazing animals, the bivalves are chiefly filter feeders, and this means that they are relatively sedentary creatures with a consequent loss of a definite head, radula and ventral foot. Their gills, on the other hand, are greatly developed to afford an effective filtering system. A few bivalves can "swim" by opening and closing their shell valves.

The third class of molluscs to be considered, the Cephalopoda, again differs strikingly from both the previous classes mentioned. The squids, octopi and cuttlefish are all carnivores and free-swimming, and consequently the molluscan shell has become lost or reduced, the body streamlined for fast movement, the sense organs very highly developed, and the mantle cavity modified for propulsion. Although very specialised for their way of life, these cephalopods are hardly recognisable as relatives of the other molluscan classes, and they emphasise the extreme genetic plasticity of the molluscan body plan. The success of this adaptability is reflected in the number of species of molluscs; 80,000; second only to the number of arthropod species.

Within the main structural framework of the chief classes of molluscs there is more adaptive radiation. Indeed at this level it is often impossible to determine true evolutionary lines because of convergence and divergence.

It would be unfair to leave a discussion of the molluscs without mentioning some of the fossil forms. Throughout the whole geological time scale, molluscs form some of the most important marker fossils, and are some of the best known and most abundant fossils as well. Of these the cephalopod fossil ammonites and belemnites are probably the most interesting, and are familiar to most students of evolution because of the gradually increasing complexity of the suture lines throughout the evolution of various ammonite lines.

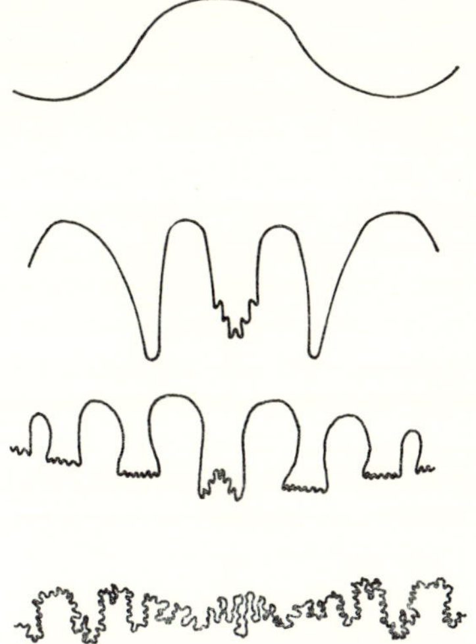

Fig 4.1. Diagram illustrating the increase in complexity of cephalopod suture lines, the simplest suture is at the top, the most advanced and complex at the bottom

The other commonly held belief about the evolution of ammonoids, however, must be treated cautiously. This is the idea that there was an orthogenetic progress from straight forms through curved and coiled, to secondarily straight forms, In fact many of these varieties co-existed, and there are very few palaeontological items of evidence which support this idea.

The Arthropoda

We have seen in the last chapter that the arthropods burst upon the Cambrian scene in the form of trilobites. These distinctive animals are now extinct, but the body plan of a firm exoskeleton plus jointed appendages and haemocoele was so successful that today arthropods comprise three-quarters of the million or so known animal species.

The fossil history of arthropods is interesting, but rather unhelpful. Thus by the Ordovician, not only the trilobites, but also ostracod crustaceans and arachnids in the form of the giant eurypterids (now extinct) and *Limulus* (the king-crab) had appeared. All the remaining arthropod classes had appeared by Devonian times, though winged orders of insects are not known before the Carboniferous. The radiation, and in particular the colonisation of the land by arthropods, paralleled that of the vertebrates, although there is palaeontological evidence to suggest that the arthropods scored "firsts" in both air-breathing and flight; in the latter case by a margin of something in the order of 100 million years. The insects themselves had radiated to produce 80 per cent of the living orders of Permian times, though an evolutionary explosion occurred in Cretaceous times, associated with the similar explosive radiation in flowering plants. For flowering plants and their pollinating insects are mutually dependent, one relying on the other for effecting its reproduction, and using as a lure the excess pollen and nectar which the other comes to require as food. It is interesting to speculate whether colour vision in insects evolved at this stage, or whether flowering plants took advantage of a pre-existing sense to make themselves more attractive to insects.

Thus the fossil history of arthropods is one of successive radiations, with no clues as to origins and affinities, and for these indications we have to turn to a study of living forms.

The one arthropodan class which stands on its own as being possibly primitive and ancestral to all other classes is the Onychophora. The features considered as being primitive are the thin cuticle, the muscular body wall, the presence of cilia, and the presence of unstriped muscle. In the past the onychophorans have been classed with the annelids, or elevated to a phylum status on their own. It has also been said of them that they form a missing link which would have had to have been invented had they not already existed. The implication in this latter statement is that *Peripatus* and its related genera indicate an evolutionary connection between the annelids and arthropods. That there is such a connection is undoubtedly true, but there are difficulties in attempting to derive the one from the other.

Manton's detailed work, already mentioned, leads her to suggest the following scheme of events:

The arthropods may well be of polyphyletic origin. This is suggested by a detailed study of mandibular mechanisms throughout the arthropod classes. One line of evolution led from the Onychophora (which Manton believes is the most primitive arthropod stock) through the myriapods to the hexapods. The reduction in leg number to six, she states, has occurred several times independently, in the Collembola, Protura, Diplura, Thysanura and Pterygota, and is associated with the advantages of faster running initially. She believes the Onychophora evolved from early annelids which possessed parapodia, but without acicula (long chaetae) to support them. *Peripatus* moves by appendicular movement only, the longitudinal muscles acting only indirectly by altering the length of the body via the hydrostatic skeleton. This alteration in length is necessary for the change in gait which accompanies the animal's change of speed. The significance of this latter fact is that the whole body changes its length, not just localised regions, and thus a non-compartmental fluid skeleton would be more efficient for this type of locomotion than a septate coelom.

There must have been strong selective pressure for the replacement of the coelom by a haemocoele, for it is clear from arthropod embryology that this is what must have happened, and there are two theories as to how it happened:

1. The haemocoele arose by peripheral parts of the blood system gradually expanding and obliterating the coelom.
2. The coelom was lost following the development of a rigid exoskeleton, and the haemocoele developed when the need for a hydrostatic skeleton reappeared later in certain secondarily soft-cuticled arthropods.

Neither of these theories is entirely satisfactory, and the development of a haemocoele is one of the many mysteries of arthropod evolution. Manton does suggest, however, that the selective advantage of the haemocoele may be that it does not leak to the exterior and cannot rupture capillaries when transmitting high internal hydrostatic pressures.

Further Reading

Borradaile, Eastham, Potts and Saunders. (1961) *The Invertebrata.* Cambridge University Press.

Clark, R. B. (1964) *Dynamics in Metazoan Evolution.* Oxford University Press.

Garstang, W. (1951) *Larval Forms.* Oxford University Press.

Lemche, H. (1957) *Nature,* **179** (Neopilina).

Morton, J. E. (1958) *Molluscs.* Hutchinson.

Phillips Dales, R. (1963) *Annelids.* Hutchinson

Stephenson, J. (1930) *The Oligochaeta.* Oxford University Press.

Tiegs, O. W. and Manton, S. M. (1958) "The Evolution of the Arthropoda". *Biol. Rev.* **33**.

5 echinoderms and the origin of vertebrates

The echinoderms are a distinctive and well-known group of animals. Every child would recognise a starfish, and many people would know a sea-urchin, though fewer would know the crinoids or holothuroids. The main diagnostic features of the phylum are the fundamentally radial symmetry (often pentamerous) and the presence of protrudable tube-feet. Also as we have seen, members of the phylum possess a coelom which develops enterocoelically, a feature which allies them to the chordates.

Crinoidea

The most primitive echinoderms known are the sea-lilies and feather-stars. These are the earliest representatives of the phylum known from the fossil record, and they have persisted almost unchanged to the present day. They are sessile forms, with a stem of varying length which bears on top a theca, from which the arms project.

The earliest true crinoid known is *Dendrocrinus* from the Lower Ordovician, but antecedent forms known as eocrinoids have been found in the Lower Cambrian.

Asteroidea

The familiar starfishes consist of a flattened central body area with a number of projecting arms, though there is no sharp delimitation between the arms and the rest of the body. Asteroids are all free-living and possess a ventral mouth and a dorsal anus. The earliest fossil form known is *Chinianaster* from the Lower Ordovician of France.

Ophiuroidea

The brittle-stars differ from the members of the Asteroidea in having the central disc of the body relatively small and sharply delimited from

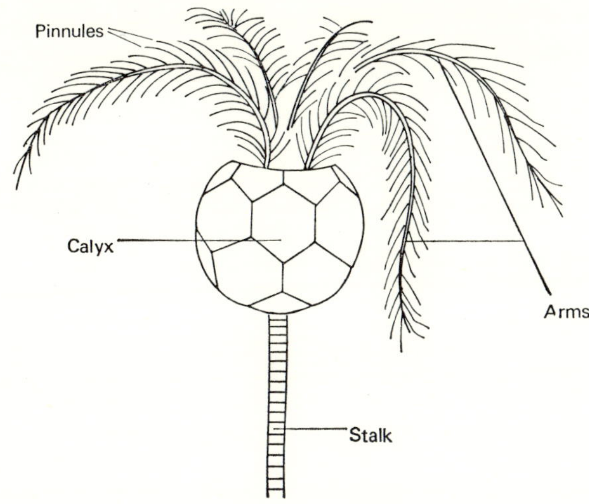

Fig 5.1. A crinoid

the arms. As the common name implies these arms are extemely fragile. Their evolutionary affinities are extremely interesting. Superficially, they appear to resemble the starfishes rather more than any

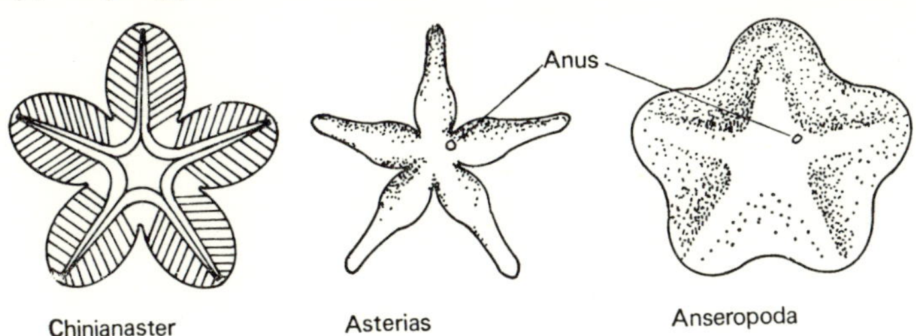

Fig 5.2. Various asteroids

other group of echinoderms, but on closer examination, their anatomy, physiology and early larval stages suggest affinities with the echinoids. There is also, however, much fossil evidence available, which indicates clearly that the asteroids and ophiuroids are closely related. It is quite probable that *Chinianaster* was ancestral to both these classes.

Echinoidea

The sea-urchins are the only echinoderms having a rigid calcareous test. Primitively they are radially symmetrical, but many echinoids have developed a definite bilateral symmetry, which means that they move in one direction only. The earliest fossil echinoids are found in

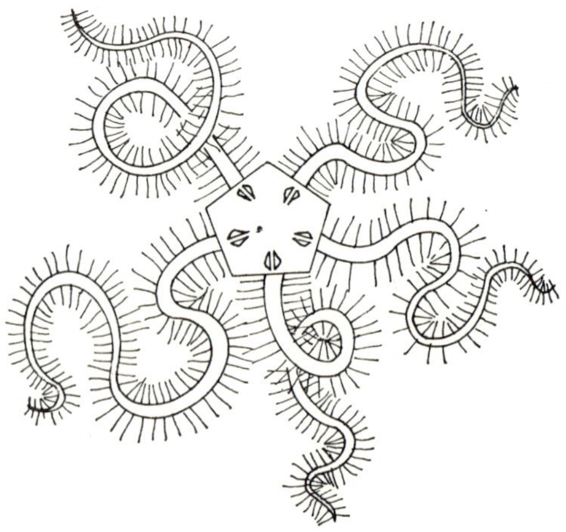

Fig 5.3. *Ophiocomina*, an ophiuroid

the Ordovician (e.g. *Bothriocidaris*), but their origins and even the early course of evolution within the group are obscure. The first

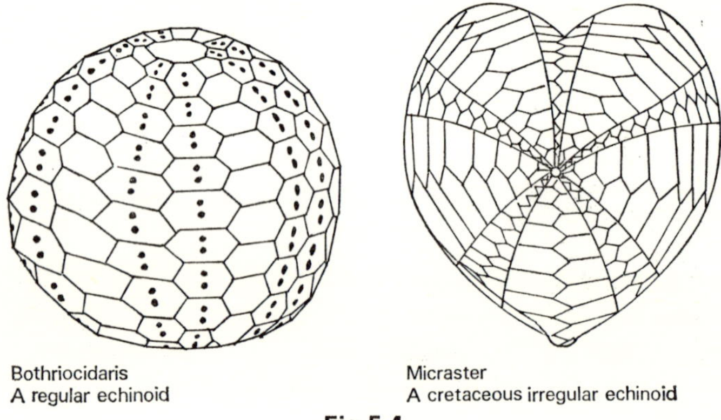

Bothriocidaris
A regular echinoid

Micraster
A cretaceous irregular echinoid

Fig 5.4.

bilaterally symmetrical irregular echinoids appeared in the early Jurassic, and later become so abundant as to act as important marker fossils.

Holothuroidea

The sea-cucumbers also move mostly uni-directionally, but they have changed the orientation of their body axis instead of becoming bilaterally symmetrical. At the anterior end of the body is the mouth, surrounded by a ring of tentacles, which are actually much branched tube-feet. Respiration is carried on in internally branching tubes, known as respiratory trees, in and out of which water is alternately pumped. These trees are best known, however, from the fact that they can be everted through the anus, together with some special Cuvierian Organs, as a method of defence. As the holothurians have a soft, but tough skin, there is a great paucity of fossil evidence compared with the other classes of echinoderms, which have much better developed skeletal structures. All that can be said is that they had evolved by the Devonian, where they first definitely make their appearance.

Relationships within the Echinoderms

It is extremely difficult to form any certain ideas as to the evolution of the group. That the crinoids are the most primitive of the living echinoderms is widely accepted, as not only do they have a sessile habit, but they appear much earlier in the fossil record than the other classes. Of the origins of the other classes very little can be said. We have seen that the Asteroidea and Ophiuroidea are probably descended from common ancestors in the not-too-distant past, but all that can be said of the echinoids and holothuroids is that they must have arisen from the main line of the free-living echinoderms at some point before the separation of the asteroids and ophiuroids.

Echinoderm Larvae

The larval stages of the echinoderms are interesting both in their own right, and more particularly because of their probable connection with the origin of the chordates. Although all adult members of the phylum are based on a radially symmetrical plan, the larval stages show a genuine bilateral symmetry, though considerable variation occurs between the different classes even at this stage of development.

The basic type of echinoderm larva is known as the Dipleurula. In most forms which have larvae, this is the first stage which develops, although it may become modified later. The dipleurula has a single band of cilia which loops in front of the mouth and in front of the anus as shown in the diagram.

The simplest modification of the dipleurula is found in the holothuroids, where the ciliated band develops a wavy appearance, and justifies the giving of a new name to this larva, the Auricularia (Fig. 5·7).

Later, the band becomes even more sinuous, and ultimately breaks up into a series of rings round the body, at which point the larva is now known as a Doliolaria (Fig. 5·7).

In the Asteroids, the larval development takes another course with the formation of a Bipinnaria larva. Here the dipleurula ciliated band

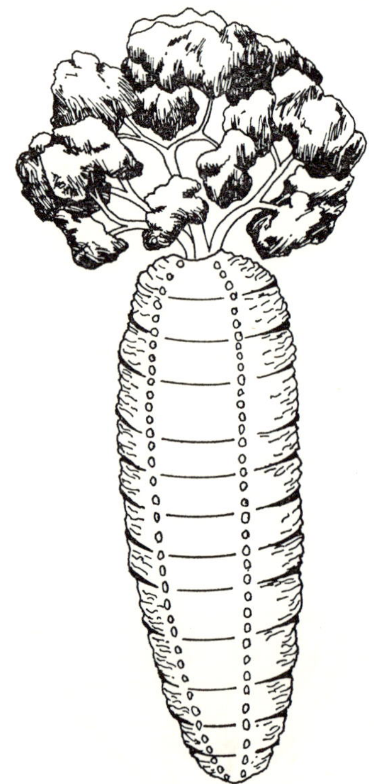

Fig 5.5. *Cucumaria,* a present-day holothuroid

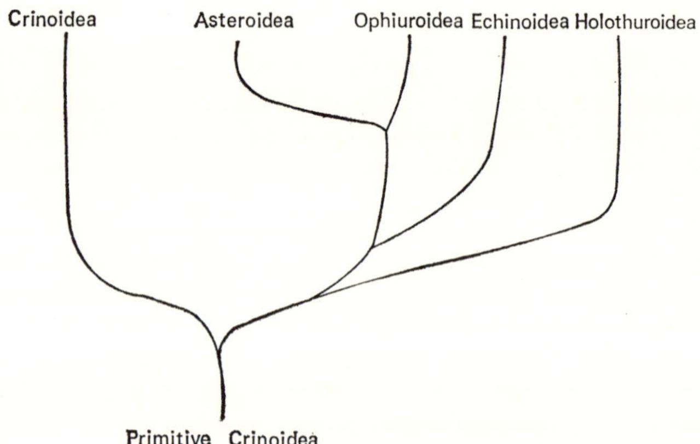

Fig 5.6. Evolution of echinoderms

becomes split into two separate bands, one in front of, and one behind the mouth (Fig. 5.7).

Yet another larval form is developed in the echinoids and ophiuroids. This is the Pluteus, which develops via the dipleurula and auricularia stages. Four larval "arms" are then pushed out, and the ciliated band follows the outline of these (Fig. 5.7).

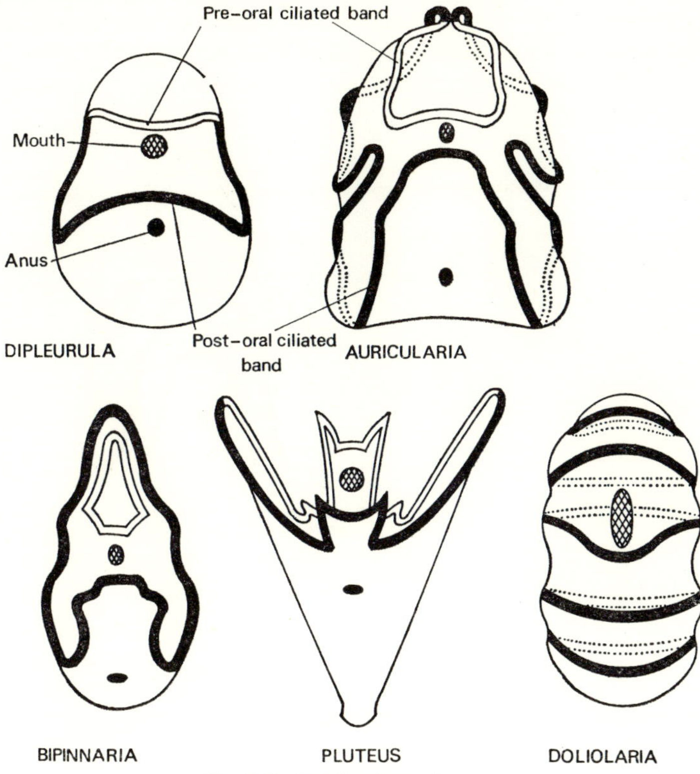

Fig 5.7. Echinoderm larvae

Other larval forms develop, and are very characteristic of certain echinoderm groups, but most of these can be derived from one of the forms described above.

Invertebrate Chordate Classes

Before considering the likely evolutionary relationships between the echinoderms and the chordates it is necessary to review the probably unfamiliar invertebrate chordate classes, as it is in these and the echinoderm larvae that the clues are found to their probable evolutionary affinities.

The invertebrate chordates (or Protochordata) are typified by possessing all the chordate characteristics at some stage in their life-history except for a vertebral column. There are three main subphyla:

1. The Hemichordata

These are chordates with a nerve cord restricted to the neck region and a small notochord arising from the pharynx. The Enteropneust hemichordates such as *Balanoglossus*, are worm-like, free-living, filter-feeding animals. Their main point of interest lies in the fact that they

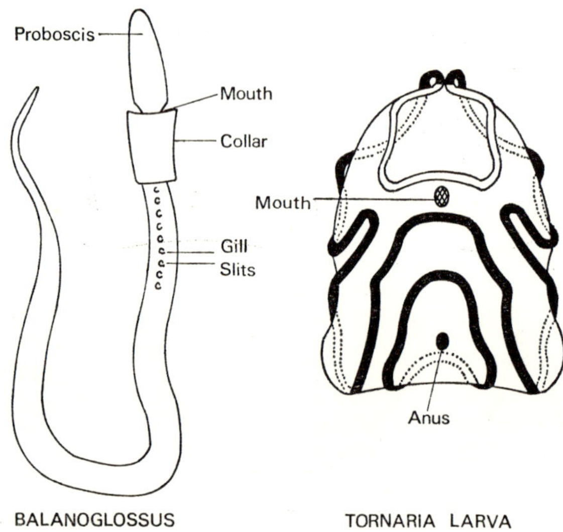

Fig 5.8.

possess a pelagic larval stage known as a tornaria, which is very similar to the auricularia larva of holothurian echinoderms. The tornaria differs only in the presence of an extra band of cilia around the anus.

The second group of hemichordates, the Pterobranchiata, consists of two or three genera of small tube-dwelling forms such as *Rhabdopleura* and *Cephalodiscus*. They have larval stages resembling the cyphonautes larva of the Ectoproct Polyzoa, and are probably fairly closely related to the extinct graptolites.

2. The Urochordata

This group includes the most atypical chordates known. Thus in many cases the adult stage possesses no chordate features apart from the presence of visceral clefts, and it is only in the larval stages that a nerve cord and notochord are found. The ascidians are the urochordates (or tunicates) best known in Britain. *Ciona* is a typical example, and consists as an adult of a sessile, filter-feeding, sac-like structure. The body is enclosed in a "tunic" of a cellulose-like substance, which is a characteristic feature of all the members of this subphylum.

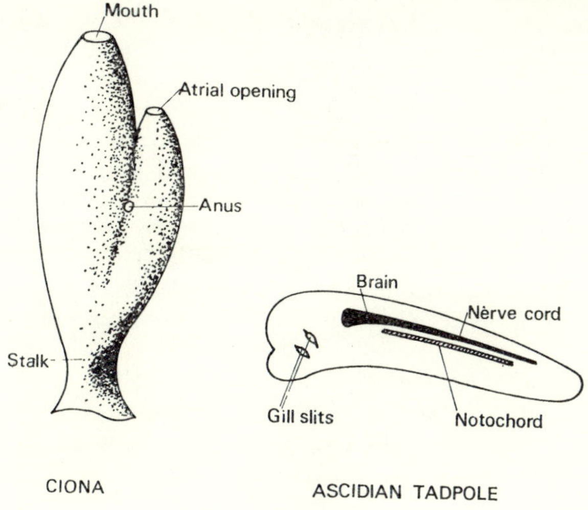

Fig 5.9.

It is only when one considers the ascidian larva, however, that one is aware of the features that enable the animal to be classified as a chordate. The free-swimming tadpole-like larva, which is about half a centimetre long, has both a notochord and a dorsal hollow nerve cord

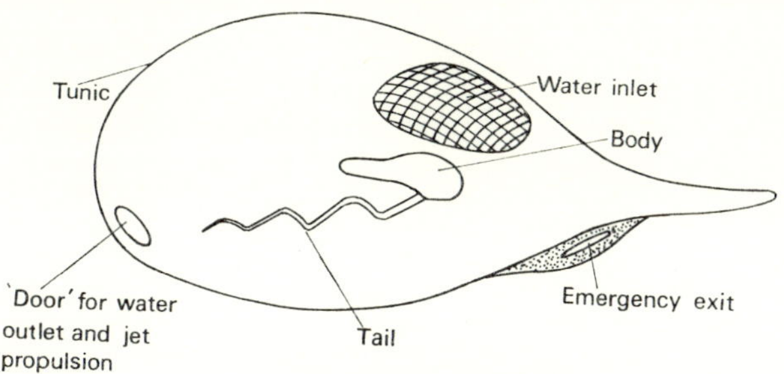

Fig 5.10. ***Oikopleura***

in its tail. It also possesses gill slits opening from the pharynx to the exterior. All these typical chordate characters vanish, however, on metamorphosis except for the gill clefts, which become greatly subdivided to form the basket-work pharyngeal wall of the adult ascidian.

The second main group of urochordates is found in the class Larvacea or Appendicularia and is typified by the genus *Oikopleura*. This tunicate leads a completely pelagic life, and the sexually mature adult retains the body structure of the larva. Its tunic is not attached to the animal, but forms a "house" surrounding *Oikopleura*, from which the latter can escape if necessary. The tail is of interest because it is normal in structure, but projects at right angles from the ventral surface of the body half-way along.

The Thaliacea comprise the third class of urochordates. There are three orders within the class, but all are free-swimming pelagic forms, and all move by contracting circular bands of muscles, which results in water being squirted out directionally and hence propelling the animals forward. The Pyrosomids (Order 1) are colonial and have no larval stage. They are notable for their ability to phosphoresce, due to the presence of luminescent symbiotic bacteria. The Salpids (Order 2) also lack a larval stage, but the Doliolids (Order 3) possess a typical tailed ascidian larva which resembles *Oikopleura* in several respects.

3. The Cephalochordates

This subphylum includes the protochordates most easily recognisable as chordates. *Amphioxus* is the most well-known member of this group,

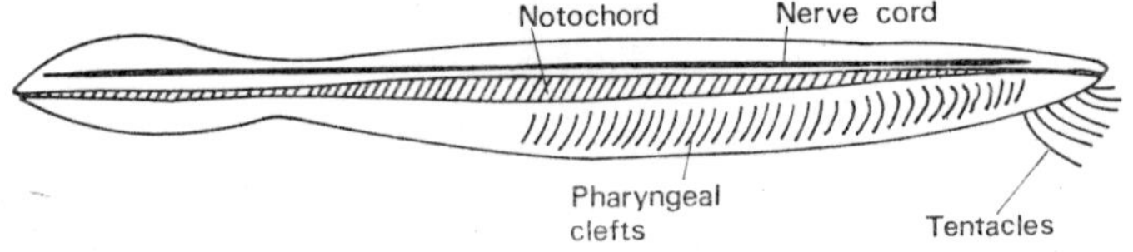

Fig 5.11. ***Amphioxus***

and possesses the diagnostic feature of the subphylum; a notochord which runs the full length of the animal in both the larval and adult stages. There is, however, no trace of a vertebral column.

Other typical chordate features, such as the gill slits can be seen from the diagram.

Chordate Evolution

The chordates are so successful a group and possess such a distinctive body organisation that speculation as to their ancestry was bound to be a major point of interest to the early evolutionists. Before the affinities of the protochordates were understood, however, this speculation was bound to be somewhat wild. Thus it was proposed very seriously on several occasions that the chordates were derived from the coelenterates, or from the annelids, or from the arachnid arthropods. The extraordinary anatomical contortions that these evolutionary steps would require, however, are so obvious to anyone with even a rudimentary knowledge of the organisation of these phyla, that details of the straightforward biological objections to such proposed ancestries will not be expounded here.

Garstang was the first to outline a reasonable theory of the origin of vertebrates, although stated baldly, his ideas sound even more outrageous than those mentioned above, For Garstang placed the Echinoderms as the group immediately ancestral to the chordates.

Garstang's Theory

The free-swimming tadpole of the ascidian urochordates is the starting point for Garstang's hypothesis. He supposed that this evolved into a free-swimming fish-like chordate, which eliminated the sessile adult stage and became capable of sexual reproduction (Paedogenesis). He assumed that *Balanoglossus* and *Amphioxus*, both being free-living in the adult form, lost their sedentary habit, retained their larval bilateral symmetry, and adopted a burrowing existence. The Larvacea (e.g. *Oikopleura*) he derived neotenously via the Thaliacea.

Extrapolating backwards, Garstang noted the similarity that we have already seen between the tornaria larva of the enteropneusts and the echinoderm auricularia. By an ingenious piece of imaginative reconstruction, he derived the ascidian larva from an auricularia by pushing the circumoral ciliated band dorsally and curling it inwards together with its associated nervous tissue to form a dorsal hollow nerve cord. The pre-ascidian sedentary adult stage he assumed to be tentacle or lophophore feeding. Tail muscles and notochord, Garstang assumed, arose *de novo* to assist with the locomotive problems facing a gradually enlarging free-swimming fish.

Later Theories

Subsequent theories of chordate origin are basically modifications of Garstang's ideas. Thus his framework of evolution from the echinoderms via the protochordates has remained basically unchanged. Berrill, however, in his book, *The Origin of Vertebrates*, does challenge Garstang on two issues. Firstly, he believes that the ascidian tadpole is a new invention of that group, and secondly, having refuted the auricularia origin of the tadpole, he also discards the tadpole as a chordate ancestor, arguing that it is a short-range site-selector with a very limited life. Berrill then seeks to derive the chordates from the other pelagic adult protochordates, such as doliolids, salps and larvaceans. He claims that the conditions prevailing at this time, of numerous slow-flowing rivers depositing a rich burden of debris and detritus into shallow seas, provided the incentive for the protovertebrates to move gradually upstream into the parts of the rivers richest in food. There would thus have been great selective pressure for the development of a strongly swimming animal, well endowed with directive sense-organs. Berrill then postulates that the rising land and consequently faster flowing streams produced a divergence in the evolution of the proto-vertebrate stock. Those members of the population that were strong swimmers and were producing large yolked eggs, also developed osmo-regulatory powers and moved upstream to produce the first vertebrate stock. The rest of the population, unable to move against the stronger currents, assumed a sedentary life on the sea-bed. This stock produced the present-day Acraniates.

Carter, in reviewing Berrill's ideas, propounds the idea of a free-swimming ancestor, radiating to produce all present-day chordates, sessile and free-swimming.

Whitear and Bone have subsequently put forward alternative theories, also based on ultimate development from echinoderm larvae, but it is not possible to go into them in full detail here, as a critical appraisal of the ideas requires a detailed understanding of the anatomy and embryology of the lower chordates, which is outside the scope of this book. The topic is one of tremendous interest, however, and it appears likely that in the fairly near future, some more positive ideas of affinities will be gained from biochemical studies.

Further Reading

Berrill, N. J. (1955) *The Origin of Vertebrates*. Oxford.

Bone, Q. (1960) "Origin of Chordates from Hemichordates". *J. Linn. Soc. (Zool.)*, **44**, 252-69.

Carter, G. S. (1957) "Chordate Phylogeny". *Syst. Zool.* **6**, 187-92.

Clark, A. M. (1962) *Starfishes and their Relations*. London.

Fell, H. B. (1948) "Echinoderm Embryology and the Origin of Chordates". *Biol. Rev.* **23**, 81-107.

Garstang, W. (1928) "The Morphology of the Tunicata and its bearing on the Phylogeny of the Chordata". *Quart. J. Micr. Sci.* **72**, 51-187.

Harvey, L. A. (1961) "New Speculation on the Origin of the Chordates". *Sci. Prog.* **49**, 507-13.

Kerkut, G. A. (1961) *Implications of Evolution*. London.

Nichols, D. (1966) *Echinoderms*. London.

Whitear, M. (1957) "Some Remarks on the Ascidian Affinities of Vertebrates". *Ann. Mag. Nat. Hist.* **10**, 338.

6
fishes and amphibians

As all students of zoology are well aware, the word "fish" is not very helpful in any scheme of classification, except possibly as a term to include all those vertebrates that are not tetrapods. The range of organisation found within the group normally described as fish is truly great, ranging from the jawless parasitic lampreys to the most advanced teleosts, although it is safe to say that all fish are more primitive evolutionarily than amphibians.

The advance from invertebrate to vertebrate is a huge one structurally, and as we have seen, the fossil record gives us no clue at all as to the intermediate stages. All the ideas as to the echinoderm and acraniate course of evolution are purely speculative and based on the anatomy, physiology and embryology of living forms. The only fact which is certain is that the vertebral column arose as a method of mechanically strengthening the notochord, whilst maintaining its flexibility.

The fossil record opens in the Ordovician as far as vertebrates are concerned. Fragments including undoubted fish scales are known from fresh-water sediments, which implies that the early evolution of vertebrates took place in fresh water, and further explains the fact that no trace of vertebrates is known from pre-Ordovician strata, which are all essentially marine deposits. The first vertebrate fossil which shows any detail is a single impression from the early Silurian of England known as *Jamoytius*. Although difficult to interpret, it shows a notochord, regular myotomes and continuous lateral fin folds. It has been described as being reminiscent of *Amphioxus* in some respects.

Fossil vertebrates start to appear abundantly in the late Silurian and early Devonian, and are known collectively as Ostracoderms, because of the heavy armour plating that characterises most of them. They were, however, very obviously specialised even at this early stage, and there were six different orders which must have been the product of a considerable evolutionary history. Structurally the Ostracoderms were very interesting, for they were all jawless, and like the living agnathans, lampreys and hagfishes, had no paired fins, and only two semi-circular canals. *Cephalaspis* is an example of the best known group of ostracoderms. It was a flattened, bottom-living form with a dense dermal armour of true bone. The head showed a pineal opening, ten gill-slits concerned with filter feeding, and some peculiar areas which

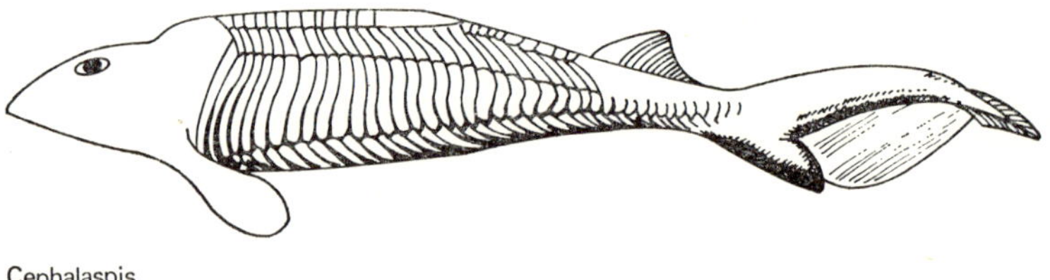

Cephalaspis

Fig 6.1.

it has been suggested were electric organs. The presence of the dense armour might appear unnecessary at first, but it must be remembered that these early vertebrates were contemporaries of the giant eurypterids.

There was a tendency amongst one group of ostracoderms to a reduction in armour and the adoption of a free-swimming habit. This group is exemplified by *Birkenia*, which had a reversed heterocercal tail (in contrast to the heterocercal and supposedly primitive tail of

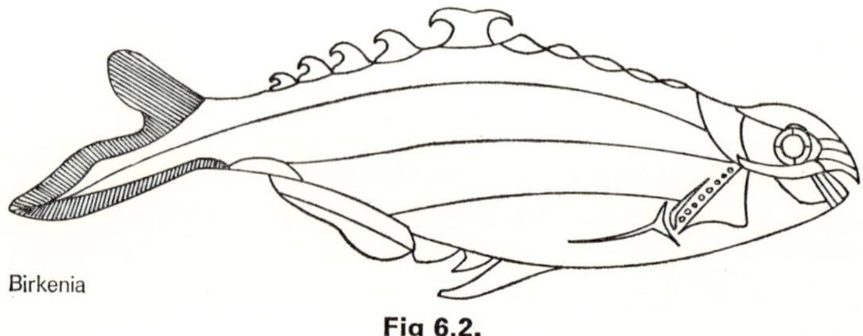

Birkenia

Fig 6.2.

Cephalaspis) to enable it to swim up through the water. Although its body was laterally compressed, it must have been very difficult for it to control its swimming without paired fins, which are normally necessary to prevent rolling, and to control pitching and yawing.

The ostracoderms were undeniably successful in the niches they occupied, but they became extinct by the end of the Devonian due to increasing competition from the rapidly evolving jawed fishes. All these bony agnathans were specialised and their only surviving relatives, the semi-parasitic lampreys and hagfishes, are also undoubtedly specialised, so although they are interesting as representing a stage in the evolution of vertebrates, none of them can be expected to indicate what the direct jawed vertebrate (gnathostome) ancestor was like. Indeed, the gnathostomes may well have originated as far back as the Cambrian era, though the fossil evidence as we have seen, is non-existent.

The Advent of Jaws

The origin of jaws marks one of the major evolutionary steps on the road to the higher vertebrates, for it is doubtful whether there was much more evolutionary potential remaining in the jawless vertebrates, and they had probably exploited all possible niches in their environment.

The early vertebrates probably had a large number of gill slits, each supported by a branchial arch of bone or cartilage. We have seen that there were ten gill slits in some ostracoderms, but embryological considerations indicate that this number had been reduced to seven in the immediate gnathostome ancestors. Of the remaining branchial arches, the first pair lost their function as gill supports, and hinged forward to form jaws. Also, as is shown by all living gnathostomes, the second branchial arch also became modified to assist in the jaw suspension at first and the remaining first gill slit became correspondingly reduced to form a respiratorily functionless spiracle.

As might be theoretically expected, there must have been a stage in vertebrate evolution when the spiracle still functioned as a gill slit and the hyoid arch was unattached to the newly modified mandibular arch. This evolutionary stage is represented by the only class of vertebrates which has become completely extinct, the placoderms. These first gnathostomes appeared in the Upper Silurian, flourished throughout the Devonian, and finally became extinct in the early Permian. There were at least six different lines of placoderms evolving during this time, but they all died out completely, and they all possessed a bony skeleton and paired fins in addition to their jaws. Apart from these features they had little in common, and ranged from the small unspecialised acanthodians, such as *Climatius*, to the giant arthrodire predators, like *Dinichthys*, which reached lengths of 9 metres.

Chondrichthyes

The ostracoderms and placoderms each had their day and then either became extinct or reduced to representation by a few specialised survivors. Neither class was ancestral to our living fishes, but both classes of living jawed fishes, the Chondrichthyes and the Osteichthyes must have passed through ostracoderm and placoderm stages in their evolution.

The Chondrichthyes, or cartilaginous fishes, represented by living sharks, rays, and dogfishes, are often considered to be primitive relative to living teleosts, but as we have seen, bone appeared very early in the evolution of vertebrates, and the Osteichthyes are actually known from the fossil record at an earlier date than the shark ancestors.

The other features typical of the chondrichthians are that they have internal fertilisation, separate external gill openings, a heterocercal tail and no swim bladder. The earliest cartilaginous fish are known from Upper Devonian strata and are superficially very similar to modern sharks. Some of these forms, however, such as *Cladoselache*, are preserved in such fine sediments that details of their internal anatomy are known, and these make it clear that despite superficial resemblances, they represent an evolutionary line that could not have led to modern sharks. Altogether four main lines of chondrichthian evolution are known, and it is interesting that three of these were found exclusively in marine deposits. It is therefore clear that at an early stage the shark ancestors left fresh water for the sea, and there evolved out of competition with the ostracoderms, placoderms and early bony fishes.

Two of the four stocks of early chondrichthians became extinct. These were the fresh water group, the Pleuracanthodii, and the Cladoselachii. Of the remaining lines, one diverged to give the modern sharks and dogfishes on one hand, and the skates and rays on the other. The final line gave rise to the comparatively rare and aberrant modern chimaeras.

During the evolutionary history of the living cartilaginous fish, one interesting development has been in the method of jaw suspension. The earliest chondrichthians had an amphistylic suspension in which the upper jaw was articulated with the braincase both by the hyomandibular cartilage and by a separate postorbital process. This gradually gave way to the hyostylic suspension, in which the hyoid articulates with the jaw and skull, but the postorbital articulation vanishes. The advantage conferred by this was presumably an increase in jaw mobility.

Bony Fishes

The Osteichthyes form a coherent group because they not only possess a bony skeleton and bony scales, but also have other features in common such as an operculum covering all the gill slits, and a swim bladder or lung, which is of great evolutionary significance.

The bony fish first appear in fresh water sediments of the middle Devonian, a date considerably earlier than that of the appearance of the first chondrichthians, as has been already pointed out. Even at their first appearance, however, the bony fishes had already diverged into two stocks; the sub-classes Choanichthyes and Actinopterygii. More will be said about the first of these two groups later, for although the subclass is almost extinct today, their early forms were the stock from which air-breathing vertebrates evolved.

The Osteichthyes soon radiated from their fresh-water environment, and by the close of the Palaeozoic era were the dominant life forms in all bodies of water, fresh and marine.

The actinopterygians, which are the dominant group of fish today, appear from the fossil record to have been greatly outnumbered at first by their choanichthian relatives, but gradually gained until they reached numerical superiority in Carboniferous times. The earliest representatives of this group are placed in the super-order Chondrostei, and are known collectively as palaeoniscids. They were small fish, with heterocercal tails, incompletely ossified vertebrae, and functional lungs. From early unspecialised forms, these chondrosteids radiated and reached a dominant position in late Carboniferous and Permian times. During this radiation, two orders evolved which have survived to the present day. One of these is represented by the sturgeons and paddlefishes, and the other by the African genera, *Polypterus* and *Calamoichthys*.

The second stage in actinopterygian evolution is represented by the present sub-order Holostei. In this group certain trends appear which ultimately lead to the teleosts. These trends include a reduction of the upper lobe of the heterocercal tail, greater ossification of the vertebrae, loss of the spiracle, and conversion of the lung to a buoyancy-controlling swim-bladder. This group was probably largely marine, but is represented only by two fresh-water survivors, both North American; *Lepidosteus*, the gar-pike, and *Amia*, the bow-fin.

The Holostei reached their evolutionary peak in the Jurassic and early Cretaceous periods. Thereafter, they gradually declined, and were replaced by the super-order of actinopterygians which is still dominant, the Teleostei. The teleosts continued the trends observed in the Holostei. Thus the skeleton became completely ossified, the tail became symmetrical (homocercal), the scales became thinner and rounder, and the pelvic fins often migrated to a more anterior position. The success of this group of bony fish is apparent today in the oceans of the world, where living representatives of the teleosts range from electric eels, through sea-horses and flying-fishes, to flounders and angler-fish.

The story of "fish" evolution is a fascinating one, for it demonstrates just how intense selective pressures to produce an animal better adapted to its environment have been. Thus the advent of jaws must have conferred an enormous feeding advantage on their possessors, and somewhat later, the evolution of fins by which swimming could be accurately controlled must similarly have been to great advantage. The changes in the later stages of actinopterygian evolution were more subtle, but equally significant in evolutionary progress; and there were many changes which we have not mentioned, such as the production of vast quantities of spawn by oceanic teleosts, and many other equally important changes of which we are not aware.

From Water to Land

The conquest of the land by vertebrates was certainly one of the most significant steps in the history of life. This is necessarily a subjective opinion, but it is questionable whether any group of organisms would have reached a state as advanced evolutionarily as the birds and mammals, had the chordates never evolved out of the water. In appreciating this, it is salutary to consider the problems facing the first tentative terrestrial tetrapod.

1. For respiratory purposes oxygen must now be extracted from air rather than from solution in water.
2. Locomotion must involve moving the body by forces applied only to the substrate, rather than to the surrounding fluid medium.
3. The increased weight of the body resulting from the reduced upthrust from the surrounding medium must be supported by an extra strong and rigid skeleton.
4. Water conservation becomes necessary.
5. Sense organs must become adapted to function in the new medium.
6. Reproductive habits must be modified. External fertilisation is impossible on land.

It is interesting to note that these exact problems are faced and solved to a greater or lesser degree by all larval amphibians metamorphosing to the adult state.

Of these six problems, the first three must be solved before any excursion on to land can occur, the remainder can be overcome more gradually, so long as free water is readily available for return visits.

It is important to realise that the step from water to land was almost certainly a retreat rather than a conquest as far as the vertebrates were concerned. During Devonian times the originally abundant fresh-water pools, ponds and lakes were gradually drying up, as is demonstrated by the abundant series of sedimentary rocks of that period. This slow desiccation, which occurred over a period of several million years, was geologically and evolutionarily catastrophic, and must have applied severe selective environmental pressures to all the species of organisms existing at that time. Many simply died out, as the bodies of water became smaller and smaller, and the competition for the dissolved oxygen correspondingly became greater and greater. Those gill-respiring creatures that were less efficient at extracting oxygen from the water suffered first, but the least affected must have been the group of aquatic vertebrates that had already developed a primitive lung, namely the Choanichthyes.

These lobe-finned bony fishes diverged from the ray-finned actinopterygians in the early Devonian, and undoubtedly the major reason for their success was that they were capable of breathing atmospheric oxygen and thus supplementing the rather meagre supplies of that gas dissolved in the water they were inhabiting. Superficially the middle Devonian actinopterygians and Choanichthyes appeared similar, owing to their comparatively recent origin from a common ancestor. This can be seen in the illustrations of *Cheirolepis* and *Osteolepis* below.

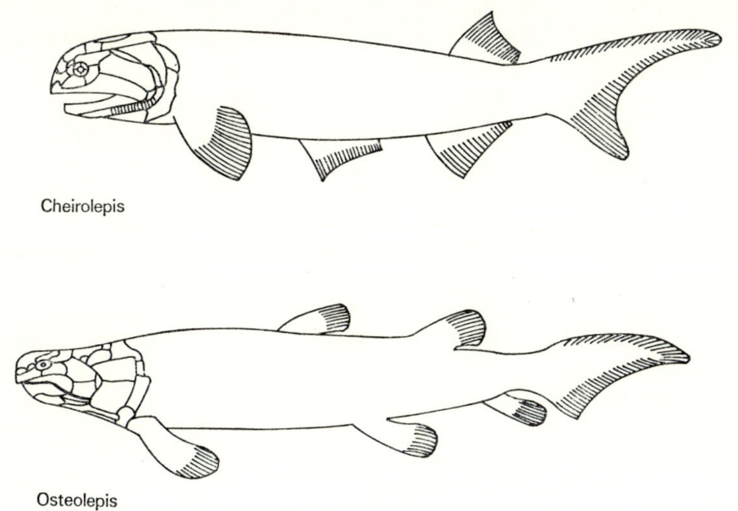

Fig 6.3.

There were, however, important differences between these two groups, and these may be summarised as follows:

	ACTINOPTERYGII	CHOANICHTHYES
1.	Heterocercal tail has no dorsal (epichordal) lobe above the body axis.	Heterocercal tail has epichordal lobe.
	Later gave homocercal tail (e.g. Teleosts).	Later gave true diphycercal tail.
2.	Paired fins supported by parallel fin rays.	Paired fins supported by median bones with radiating lesser bones.
3.	One dorsal fin.	Two dorsal fins.
4.	No internal nostrils.	Internal nostrils.
5.	Large eyes.	Smaller eyes.
6.	Ganoid scales.	Cosmoid scales.
7.	Pineal opening usually absent.	Pineal opening present.

The Choanichthyes were themselves by no means a homogenous group, however, and within them two definite orders had evolved even by the mid-Devonian. These two groups were the lungfishes or dipnoans, and the crossopterygians. The lungfishes are the better known of the two groups, as there are still three living genera. These are *Neoceradotus* in Australia, *Protopteris* in Africa, and *Lepidosiren* in South America.

The crossopterygians, despite being less well known, are far more important in being directly in the line of evolution of the tetrapods. The reasons supporting this assertion are numerous and conclusive, but the major points are as follows:

1. In the dipnoans, the internal skeleton tended to become reduced, whereas in the crossopterygians there was no such trend. This is significant because a rigid and strong internal skeleton would be necessary to withstand the extra stresses of life on land.

2. There are remarkable similarities between the skull bones of the crossopterygians and the earliest amphibians with regard to arrangement and number. Such affinities are not apparent between the skull bones of the dipnoans and the earliest amphibians.

3. Marginal teeth were lost early in dipnoan evolution, whereas they were retained by the crossopterygians (probably in association with a carnivorous mode of feeding), and in cross-section show a remarkable highly folded structure. These teeth were therefore described as labyrinthodont, and again, are extraordinarily similar to the teeth of the earliest amphibians.

4. The internal structure of the paired fins of the two groups also differed as in Fig. 6.4.

From this it is simple to see that the crossopterygian limb could have evolved into the tetrapod limb far more easily than could have the dipnoan limb.

We have so far been considering the primitive crossopterygians, as found in the early Devonian. During this period, however, two suborders evolved, one of which, the Rhipidistia, was on the line of tetrapod descent, while the other, although not so important in our immediate considerations, has achieved a vast degree of fame in recent years. This second suborder was the Coelacanthini, a group of interest only to palaeontologists until 1939, when a living coelacanth, later named *Latimeria*, was caught off the coast of South Africa. This received great press coverage as an example of a "living fossil", and it was indeed a surprise to everyone, as no post-Cretaceous remains of coelacanths had ever been found.

The First Amphibians

At what point does a rhipidistian become an amphibian? Unfortunately, there is insufficient fossil material available for this to be more than an academic question. Undoubted forms of both groups are known, but there is only one "missing link", and that is known only as one incomplete skull. It must be emphasised here that we are dealing with fossils, and thus we cannot distinguish an amphibian from a "fish" by merely observing whether or not the auricle is divided, or whether a larval stage exists. Identification must be carried out solely on the basis of the preserved hard structures. One of the most important of these criteria is whether a true pentadactyl limb is present or not. The presence of this characteristic tetrapod structure is equated with the adoption of a terrestrial life, and the earliest amphibians are identified on this criterion.

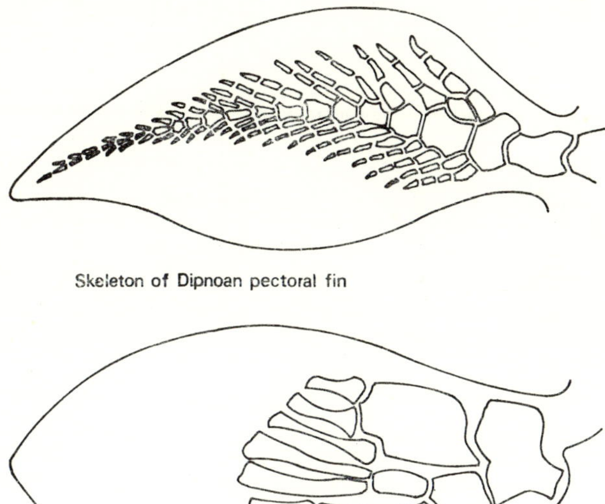

Skeleton of Dipnoan pectoral fin

Skeleton of Crossopterygian pectoral fin

Fig 6.4.

The ichthyostegids, as the earliest amphibians are known, are found in late Devonian strata in Greenland. Their skull bones, as we have already seen, were obviously derived from those of the rhipidistian crossopterygians, and the "missing link" that has been found is a skull showing the bones in an intermediate condition. In some respects, however, the ichthyostegids still resembled their crossopterygian ancestors, particularly in the possession of fin rays in the tail, and in the primitive structure of their vertebrae. The only other distinctive amphibian character that can be distinguished in these primitive tetrapods, is the presence of an otic notch at the back of the skull on each side. This signifies that an eardrum was present, and in association with this, a bone (the stapes), is found with one end inserted in the braincase, and the other end in a position that would have been attached to the tympanic membrane in life. The function of the stapes was (and is still in living amphibia) to amplify the air vibrations

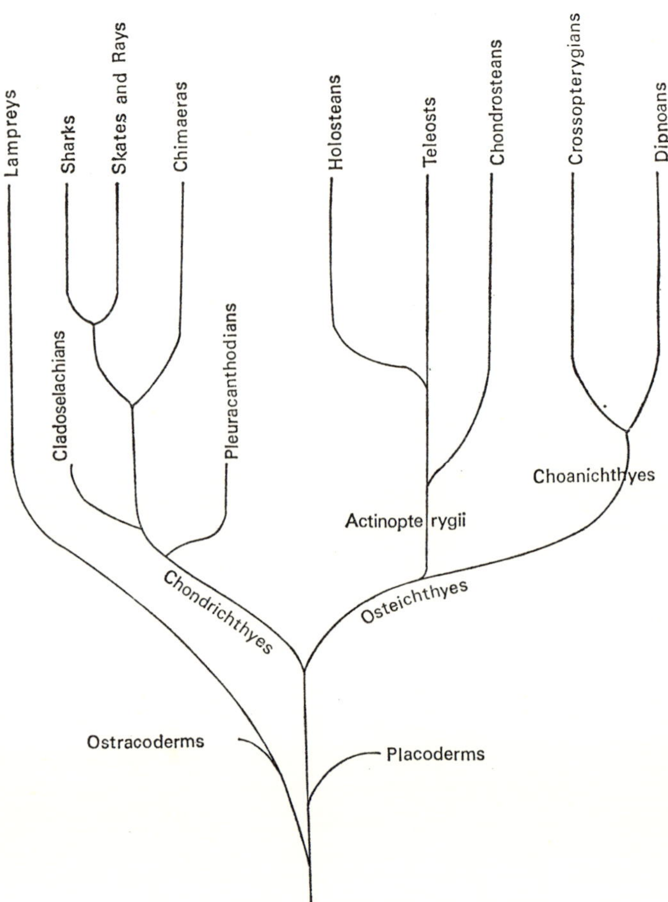

Fig 6.5. Evolutionary tree of "Fish"

impinging on the eardrum and transmit them to the inner ear. This is clearly an adaptation to life on land and the consequent necessity to receive the comparatively gentler vibrations transmitted by the new surrounding medium—the air.

The evolutionary history of the stapes is very interesting and very informative. It was originally a bone component of the second visceral arch. This later became the hyomandibula, important in fishes as part of the jaw suspension apparatus, and finally, with the adoption of the land habitat, it took on a completely new function as we have seen, but still in connection with its corresponding visceral cleft the spiracle, which in land animals becomes the eustachian tube and passage from inner to outer ear.

Amphibian Radiation

During the Carboniferous period, the primitive ichthyostegids evolved along several lines.

1. The Embolomeres

These were not advanced amphibians, but they showed a marked increase in size over their ichthyostegid ancestors. *Eogyrinus*, for example,

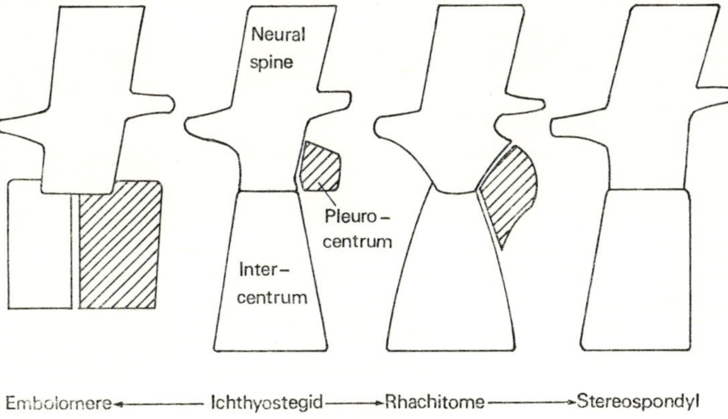

Fig 6.6.

was about 5 metres long, and although it was undoubtedly still primarily aquatic with rather weak limbs for its size, it had a well-ossified skull and a strong backbone. The nature and structure of the backbone is important in classifying these early amphibia. Basically the ichthyostegids had a vertebral centrum composed of two parts, an intercentrum and a pleurocentrum, on top of which the neural spine rested. In the embolomeres, both the inter-and pleuro-centra were present, but the vertebrae interlocked to a greater extent than had been the case in the ichthyostegids, thus giving the animals more strength and freedom of movement on land.

2. The Rhachitomes

These important amphibians evolved after the first embolomeres had appeared, but still relatively early in the Carboniferous period. They persisted until the Triassic, however, and during the Permian were among the dominant forms of life of that period. In these amphibians the two central elements were not equal, the intercentra becoming dominant and wedge-shaped. *Eryops* was a typical rhachitome, being a heavily built animal about 2 metres long, feeding on fish, and probably some small land-living tetrapods. It is also thought that these rhachitomes had gill-breathing larval stages, as remains have been found of small labyrinthodonts in various stages of growth, which although originally classified as "branchiosaurs" or "phyllospondyls", are now believed to be larval rhachitomes.

The rhachitomes, as is often the case with successful animal groups, underwent a considerable radiation themselves. Some became armoured, some returned to a fresh-water existence, and one group, the trematosaurs, invaded the sea. This latter experiment was not very successful, however, probably because they had to return to fresh water to breed, and the trematosaurs became extinct early in the Triassic. One group which evolved from the Rhachitomes became very successful, and succeeded them as the dominant amphibian group, namely:

3. The Stereospondyls

This group, which flourished and became extinct during the Triassic period, represents the final evolutionary development of the labyrinthodonts. Although very successful in their own right, they were disappointing amphibians because they returned to a completely aquatic mode of life, and in consequence of this reversion their skeleton became secondarily reduced and weakened. The pleurocentrum was entirely lost, and the vertebrae lost their interlocking processes. Some of the stereospondyls also attained enormous sizes, a feat which would not have been mechanically possible on land.

Origins of Present-day Amphibia

The three orders of living amphibia are characterised by their high degree of specialisation, and by the thinness of the fossil record of their ancestry.

The anurans (frogs and toads) are specialised for jumping, and have a considerably reduced skeleton. They are thus greatly removed from their Carboniferous ancestors. Nevertheless, it is possible to trace their evolutionary history through a few isolated fossils; *Protobatrachus* of the Madagascan Trias, and *Miobatrachus* and *Amphibamus* of the Carboniferous, and to conclude that the anurans are almost certainly of labyrinthodont ancestry.

The urodeles (newts and salamanders) and the apodans (blind, limbless, burrowing tropical forms) are even more deficient in fossil ancestors than the anurans. Because of their vertebral structure, however, they are usually assumed to have arisen from a group of small Carboniferous amphibians which we have not so far mentioned. These are the lepospondyls, which have caused many palaeontological headaches because their spool-like vertebrae are totally different from the basic ichthyostegid pattern, and in addition not formed from cartilaginous precursors, as were the labyrinthodont vertebrae. They are usually considered to have diverged early from the ichthyostegid stock, but it is quite possible that they may have had a completely different aquatic ancestor, in which case both the living and the fossil amphibia would be of polyphyletic origin.

Further Reading

Colbert, E. H. (1955) *Evolution of the Vertebrates*. London.
Goodrich, E. S. (1930) *Studies on the Structure and Development of Vertebrates*. London.
Moy-Thomas, J. A. (1939) *Palaeozoic Fishes*. London.
Noble, G. K. (1931) *The Biology of the Amphibia*. New York.
Norman, J. R. (1931) *A History of Fishes*. London.
Romer, A. S. (1946) "The Early Evolution of Fishes". *Quart. Rev. Biol.* **21**, 33-69.
Romer, A. S. (1955) *The Vertebrate Body*. London.
Stensio, E. (1926) *The Cephalaspids of Great Britain*. Brit. Mus. London.
Swinton, W. E. (1958) *Fossil Amphibians and Reptiles*. Brit. Mus. London.
Watson, D. M. S. (1919 and 1925) "Evolution and Origin of Amphibia". *Phil. Trans. Roy. Soc.* **209**, and **214**.
Westoll, T. S. (1943) "The Origin of Tetrapods". *Biol. Rev.* **18**, 78.
Young, J. Z. (1950) *The Life of Vertebrates*. Oxford University Press.

7
reptiles and birds

The reptiles form one of the more spectacular animal groups evolutionarily, for who has not had his palaeontological interest stimulated by seeing reconstructions or illustrations of the dinosaurs? Nevertheless, it is not in these bizarre and diverse lines of reptile evolution that the main interest lies, although they do make a fascinating study.

The evolutionary step from amphibian to reptile was every bit as significant as the change from agnathan to gnathostome, or from fish to amphibian. In fact it followed very closely on the latter progression, and was probably guided by the same environmental influences, resulting in a further step in the vertebrate migration from water to land.

As we saw in the last chapter, amphibians were still limited by the need to return to water for reproductive purposes, and by the danger of excessive dehydration through their skins. Reptiles have contrived to solve both these problems, although they still share the amphibian disadvantage of being cold-blooded. Inability to control body temperature is not important when the animal is living in a medium of fairly high specific heat capacity, such as water, where the maximum temperature variation is kept small, but temperature variations on land are such that death will occur readily as a result of overheating or freezing unless direct or indirect methods of temperature control are employed.

The initial evolution of reptiles was in all probability a continued "defensive" evolution, adapting to the gradual dehydration of the Devonian pools. Only later did the reptilian explosion occur, as these vertebrates started to realise the potential of land existence.

It will help at this stage to consider those features distinguishing reptiles from their amphibian ancestors, and although there are numerous minor differences, two stand out as being of fundamental importance:

1. The Amniote Egg. The development of the amniote egg must

stand as the main distinguishing criterion between amphibians and reptiles. This permitted reproduction to occur without recourse to water, and therefore allowed vertebrate life to spread further than the zone immediately surrounding bodies of water. Firstly a tough shell was laid down round the egg which prevented dehydration. Secondly, large supplies of yolk were provided which allowed the young reptile to emerge from its egg as a fairly large and active animal. And thirdly, a system of membranes inside the egg provided both a miniature "pond' in which the embryo could develop, and a storage sac for toxic excretory products.

Along with the development of the amniote egg went the elimination of a gill-breathing larval stage. Reptiles emerge from their eggs as young adults, and no metamorphosis occurs. A further corollary of this process is the necessity for internal fertilisation. Gametes can no longer be freely released into the water, and fertilisation must be effected before the relatively impermeable shell is deposited round the egg. Thus copulatory organs also had to be developed.

2. Scales. The now permanently land-dwelling reptiles were in severe danger of desiccation, and the development of scales or scutes greatly reduced the amount of water loss through the skin.

Despite these features, the earliest reptiles were still largely aquatic, and the amniote egg was originally almost certainly a precaution against drying up rather than an adaptation to a terrestrial life. Hence scales, which are now a diagnostic feature of reptiles, were not markedly present in the earliest reptilian forms.

Cotylosaurs

If we decide that the first true reptiles were the first vertebrates to lay amniote eggs, we are in great difficulty over identifying the first reptiles. Eggs themselves are only rarely preserved, and a positive association of a fossil organism with its own eggs is almost a unique occurrence. Thus the identification of the earliest reptile remains has to be done purely on osteological criteria. On this basis, the fossil *Seymouria* has for a long time been considered to be a connecting link between amphibians and reptiles. *Seymouria* was a small (60 centimetres long) tetrapod, whose fossil remains have been found in the Lower Permian deposits of Texas. It existed far too late in time to be an actual reptile ancestor, but it was probably an only slightly modified descendent of the stock which ultimately gave rise to the undoubted reptiles. There is a great deal of controversy as to the actual systematic position of *Seymouria*, because it does show a balanced mixture of amphibian and reptilian characteristics.

Amphibian features include the arrangement of the skull bones, the presence of an otic notch (usually lost in reptiles) labyrinthodont teeth, the short neck and the traces of lateral line canals.

Reptilian features were the outwardly directed orbits, the differentiated atlas and axis vertebrae, the fusion of a second vertebra into the sacrum, and finally the arrangement of the bones in the five digits in the characteristic reptilian pattern. The only way of determining the true affinities of *Seymouria* would be to obtain some knowledge of its egg-laying habits. This, however, is an extremely unlikely occurrence. Irrespective of its exact taxonomic position, *Seymouria* demonstrates beautifully the fact that the earliest reptiles evolved from the labyrinthodont stock of amphibians.

The first undoubted reptiles appeared contemporaneously with *Seymouria*. These were forms such as *Limnoscelis*, *Captorhinus*, and *Diadectes*. Although the cotylosaurs are sometimes called "stem-reptiles", the name given to the order is indicative of the cup-shaped vertebrae they possessed. These early reptiles must have walked rather awkwardly with their limbs projecting out sideways, but nevertheless, they attained lengths of up to three metres.

Classification of Reptiles

Reptilian classification is rather unsatisfactory, as the arrangement into sub-classes, based as it is purely on details of holes in the skull, is empirical, and therefore may not indicate true phylogenetic relationships. In the absence of other satisfactory classifications, however, it is necessary to adhere to the generally accepted arrangement, and accept that the taxonomic groups may be artificial.

Subclass Anapsida

This group of reptiles is characterised by having the skull roof perforated only five times; two nares, two orbits and a pineal opening comprise these basic fenestrations as they are known (Fig. 7.1). All cotylosaurs had skulls of this type. The only other reptile order to be included in this subclass is the Chelonia, which comprises the tortoises, turtles and terrapins, a very specialised group of reptiles which has remained practically unchanged since the Triassic period. That they almost certainly evolved from a Cotylosaur stock is shown by the discovery of several intermediate forms, in particular, *Eunotosaurus* of the Upper Permian.

Subclass Synapsida

In these reptiles, an extra opening had appeared on both sides of the skull, bounded above by the postorbital squamosal bones (Fig. 7.1). The function of these new openings and also of all the other fenestrations to be described, was to allow room for the bulging of the more powerful jaw muscles. Two important orders of reptiles had synapsid skulls; the Pelycosauria, and the Therapsida. The pelycosaurs, found in the Lower Permian, were often large reptiles, over 3 metres long,

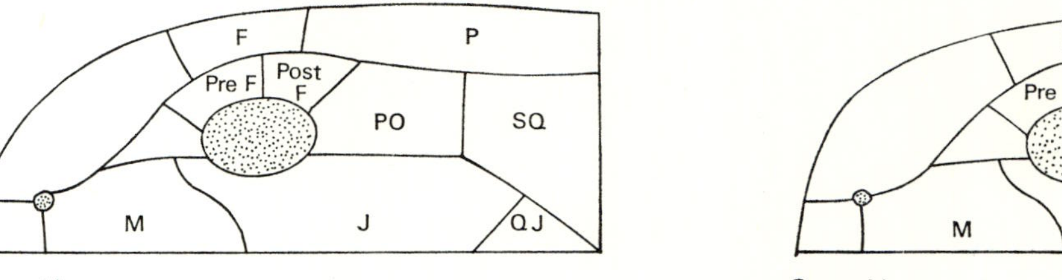

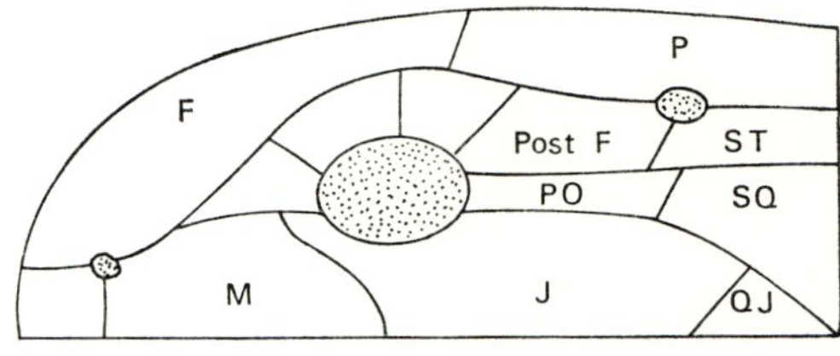

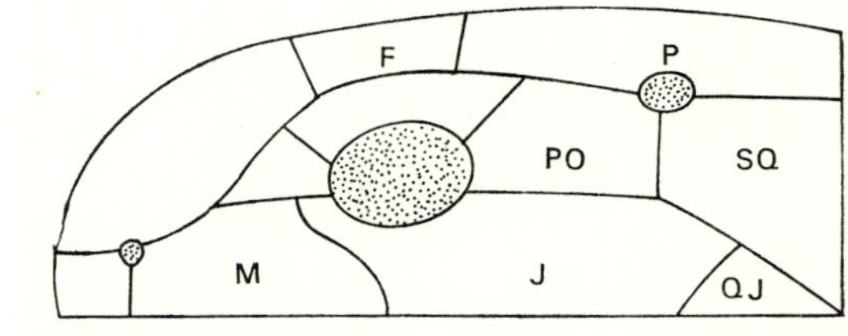

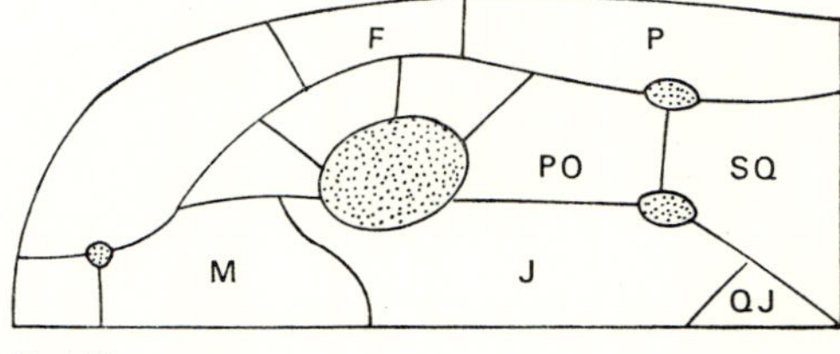

Fig 7.1. Reptilian skulls

F—frontal
J—jugal
M—maxilla
P—parietal
PO—postorbital
Post F—postfrontal
Pre F—prefrontal
QJ—quadratojugal
SQ—squamosal

but were still primitive in their gait, being belly-walkers. Some of them possessed large canine-type teeth, and others were notable for the remarkable length of their neural spines (often a metre or more in *Dimetrodon*), which must have supported a web of skin, like an inverted keel, on the animals' back. Various suggestions have been made as to the functional significance of this keel, the most likely one of which is that it served as a crude temperature regulator. Pelycosaurs closely related to *Dimetrodon* almost certainly gave rise to the other main group of synapsid reptiles, namely the Therapsida. These animals are found in sediments ranging from late Permian to the end of the Triassic and although they displayed a great variety of form, the later members were very successful. Among features contributing to this success were the limbs which developed more below the body, thus making for more efficient and faster terrestrial locomotion; somewhat differentiated teeth; and a secondary palate, which facilitates breathing. Many were carnivores. Although none of these therapsids grew very large, they were an extremely important group of reptiles, as they ultimately gave rise to the mammals. More will be said about this evolutionary development later, but we must firstly consider the other groups of reptiles, which can be divided into three further subclasses.

Subclass Parapsida

This group, also known as the Ichthyopterygia, possessed a single extra temporal fenestration on each side, bounded ventrally by the postfrontal and supratemporal bones (Fig. 7.1).

The main interest of the group, however, which is shared by the Euryapsida, is that all its members were secondarily adapted to an aquatic life. These were not fresh-water paddlers, as we shall see was the case with some of the dinosaurs, but were fully marine fish-like reptiles, thus seemingly reversing the slow process which had placed vertebrates on dry land. The ichthyosaurs are very well known extinct reptiles, and only a superficial knowledge of them is necessary to appreciate their efficient adaptations to a fully aquatic life. Thus their fusiform body shape, which gradually became more streamlined during the Triassic and Jurassic; their paddle-like limbs showing hyperdactyly and hyperphalangy (increase in the number of digits and phalanges respectively); their reversed heterocercal tail and their unsupported dorsal fin, all point strongly to their adoption of a secondarily aquatic life. There is also good fossil evidence to indicate that the ichthyosaurs were ovo-viviparous, which would have overcome the difficulties produced by having to return to shore to lay the now disadvantageous amniote egg. Along with these changes went the elongation of the snout into a typical piscivores jaw, with many small teeth. Ichthyosaurs finally became extinct in the late Cretaceous.

Subclass Euryapsida (Sauropterygia)

The well-known plesiosaurs, and the less well-known nothosaurs and placodonts comprise this group, which were, as already mentioned, secondarily adapted to an aquatic existence. The best known form of plesiosaur had a long thin neck, with a wide dorso-ventrally flattened body, and thus must have propelled itself with its paddle-like limbs mainly, in contrast to the ichthyosaurs, which flexed their bodies and tails laterally in carangiform fashion. The whole group was characterised by having the single temporal opening bounded ventrally by the postorbital and squamosal bones (Fig. 7.1).

Subclass Diapsida

This subclass includes all the remaining reptiles, living or extinct, and its members are distinguished by the possession of two extra fenestrations in the skull on each side. This is true only basically, as some of those reptiles classified here have an adult skull which differs markedly from the type illustrated above, although it is clear that in all these cases the original derivation was from a Diapsid skull as opposed to any other type. *Youngina*, a member of the primitive order Eosuchia, has a typical diapsid skull (Fig. 7.1).

There are seven other orders of diapsids and since most are well known they will be considered briefly.

Order Rhynchocephalia

Sphenodon, the tuatara which inhabits a few islands off the North Coast of New Zealand, is the only living representative of this order, and appears to have existed relatively unchanged for at least 180 million years. It is of particular interest as it still retains a functional pineal eye on top of its head.

Order Squamata

The lizards and snakes have almost certainly evolved from an eosuchian ancestor, although their early fossil history is not represented by many species before the later Jurassic. This is a fascinating group, for it includes the majority of living reptiles, but it is a difficult group from which to draw systematic conclusions. Thus one could say that the evolutionary success of the lizards and snakes is largely due to their possession of scales and their excretion of uric acid (both adaptations to conserve water), but this would only be part of the story. The adaptive radiation that has occurred within each of these suborders, demonstrates that evolutionary plasticity has contributed greatly to their general success. Although the snakes (Ophidia) are of more recent origin than the lizards (Lacertilia), the original mode of life that favoured limblessness is not clear. On one hand, it seems that the serpentine condition may have been an adaptation to an aquatic existence; whereas there is equally good evidence to suggest that all snakes were once almost blind burrowing forms, and that this was the environmental pressure that reduced them to the limbless state.

The three orders mentioned above are sometimes grouped together as the Lepidosauria, whereas the remaining five orders of Diapsids are known as the Archosauria. These "ruling reptiles" differed from the lizards and snakes and their ancestors in quite a number of respects. The most notable of these were the absence of a pineal foramen and the relatively small fore-limbs and large hind-limbs, suggesting a bipedal gait.

Order Thecodontia

These were primitive Triassic bipedal archosaurs. Never more than about 5 metres long, they were covered in bony scales, and derived their name from their hollow-based teeth. Extinct by the end of the Triassic, they nevertheless had given rise to the remaining important groups of archosaurs.

Order Saurischia

These reptiles, together with the members of the following order, the Ornithischia, are popularly grouped together to form that well-known, but artificial reptilian group, the "dinosaurs". Spectacular, because

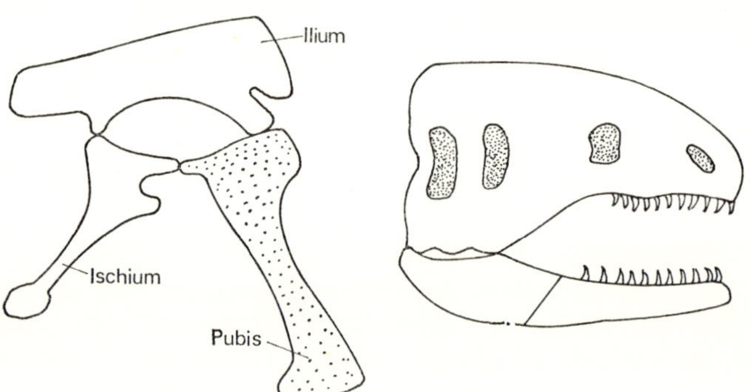

Fig 7.2. Saurischian pelvis and skull

both orders included large forms at the time when the reptiles had reached their evolutionary peak, they nevertheless became totally extinct by the close of the Mesozoic era. Amongst the many features in which the saurischians differed from the ornithischians, were the arrangement of the pelvic bones, and the structure of the skull.

The saurischians included amongst their number the giant bipedal carnivore *Tyrannosaurus*, and the even larger, though herbivorous, *Diplodocus*.

Order Ornithischia

This second "Dinosaur" order, the structure of whose pelvis and skull are shown below, were probably all herbivores, although they showed both bipedal and quadrupedal gaits, as did the saurischians.

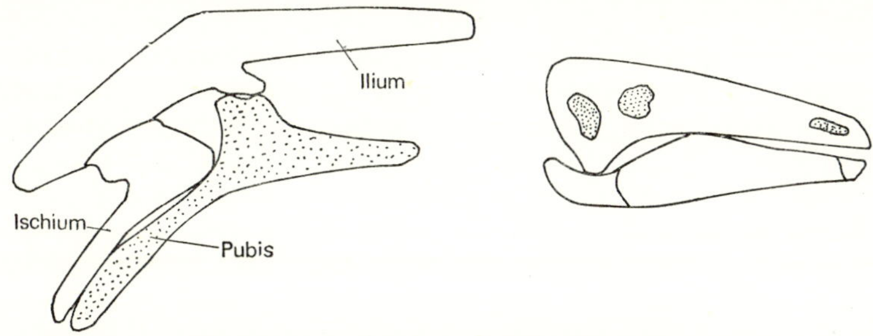

Fig 7.3. Ornithischian pelvis and skull

Iguanodon is the best known of the upright forms, and of the quadrupeds, *Stegosaurus* and *Triceratops* are both forms illustrating the great development of armour that occurred in these "dinosaurs".

Order Crocodilia

This is the only archosaur order still represented by living species. The earliest crocodiles are found in the Upper Trias. The most interesting feature of this group is that its members gradually developed a secondary "false" palate, thus paralleling the formation of a similar structure in the mammals. This development would allow them to continue breathing air, whilst their submerged mouth opened to grip and drown or eat their prey.

Order Pterosauria

With the pterosaurs, we meet the first major advance made by vertebrates since their conquest of the land, namely the adoption of flight. These aerial Diapsids flourished in the Jurassic and Cretaceous, although they became extinct at the close of the latter period, possibly as a result of competition with the more successful birds. Although there is no fossil evidence to support the theory, it seems highly likely that the ancestral pterosaurs were arboreal reptiles that developed flaps of skin stretching between their fore- and hind-limbs. These flaps would then have been used for gliding to a lower level in much the same way that flying foxes and squirrels do today. The structure of the earliest Jurassic forms, such as *Rhamphorhynchus* suggests derivation from an ancestral form such as the above. The wing membrane was supported anteriorly by a greatly elongated fourth digit (the fifth was lost) and posteriorly by the femur. This meant that there were no intermediate supporting struts, as is the case in the modern bats, and correspondingly the mechanical strength and efficiency of the wing must have been low.

Pterodactylus, a Jurassic contemporary of *Rhamphorhynchus* was about the size of a sparrow. Cretaceous pterosaurs, however, tended to reach much greater sizes, and *Pteranodon* had a wing-span approaching ten metres.

Apart from structural adaptations to flight paralleling those evolved in the birds, and described later, features of interest in the pterosaurs were that teeth were present in many species, suggesting a carnivorous or piscivorous diet; early forms such as *Rhamphorhynchus* had a long tail terminating in a flap, which probably acted as a rudder; and thirdly, the involvement of both pairs of limbs in the wing membrane or patagium, indicates that these pterosaurs must have been tree or cliff dwellers, since they would never have managed to walk on, or take off from the ground. The whole problem of pterosaur flight is cloaked in mystery, as we shall see in the next section.

More Sophisticated Flight

During Jurassic times, two distinct groups of vertebrates took to a partly aerial life. One of these groups was the pterosaurs, which we have considered above, and the other was the birds. The latter still flourish, whereas the flying reptiles were extinct by the end of the Cretaceous. What are the particular problems of flight that the birds mastered, but at which the pterosaurs were obviously not so successful?

The most apparent problem to overcome is that of gravity. Thus even the smallest vertebrates have a mass, density and body plan which will ensure that they strike the ground with a lethal terminal velocity if dropped from any appreciable height. In order to overcome this problem, some form of flight surface must be evolved, which will increase the effects of air resistance to such an extent that the velocity of descent is reduced to a safe level. Along with this must go a reduction in density if this is possible. The only way in which this can be effected is by hollowing out the dense limb bones, but only to a point where the strength of the bone is not affected, and this is often achieved by the positioning of internal reinforcing struts.

The second main point concerned with active flight, as opposed to mere gliding, is that the organism must have a power/weight ratio that will physically allow flapping flight. The anatomical factors concerned with this are relatively simple. Thus the flight surface must be able to exert a force downward on the air which is at least equal to the force due to gravity acting on the animal's body. The force that the wing can exert will depend partly on its area, and partly on the rate at which it can be flapped, but whatever the individual arrangement, the wing has to be moved by muscles. The power exerted by a muscle is, generally speaking, proportional to its cross-sectional area, and therefore flight muscles tend to be very large and to increase out of proportion to the size of the animal. Large muscles need large areas for attachment, and therefore the skeleton becomes modified, as for example in the breast bone of living birds. The upper limit for size of a flying vertebrate is decreed by the power/weight ratio. The mass of an organism increases by the cube of its increase in linear dimensions, whereas the cross-sectional area of the muscles only increases by the square of this measurement. Thus if the power/weight ratio is to remain constant as a flying vertebrate increases in size, the proportion of the total mass of the organism which is due to the flight muscles and skeletal attachments progressively increases until an obvious upper limit is reached.

Other modifications concerned with flight involve the sense organs and the metabolism. Temperature control would seem to be a "must" for organisms maintaining flight throughout all seasons of the year.

This brings us back to the pterosaurs. These reptiles possessed folds of skin serving as flight surfaces, a lightened skeleton and an enlarged sternum. But the flight muscles must have been very small, as judged by their areas of attachment, and being reptiles, the pterosaurs must have been cold-blooded. It is, of course, possible that homoiothermy developed here within the reptiles, but one would expect this to be associated with the development of some kind of insulating covering such as hair or feathers, neither of which are known to have occurred in *Pteranodon* or its relatives. It would seem from these observations that the pterosaurs must have been somewhat inefficient at flying, and may well have relied almost entirely on gliding or soaring from cliff-faces or tree-tops. It is interesting, however, that they became extinct contemporaneously with their dinosaur relatives.

The Birds

Birds have been frequently described as glorified reptiles, although one might be equally justified in applying the same epithet to mammals. They appear first in the fossil record during Jurassic times, and seem to have originated from a second reptilian excursion into the realms of flight. The success of the birds, however, as compared with the pterosaurs, was probably attributable to several factors:

1. The modification of only the fore-limbs as flight surfaces left the hind-limbs free for running, scratching, paddling and a variety of other activites. It is likely that flight developed in birds as an aid to fast running, rather than as a derivation from gliding as was probably the case with the pterosaurs.
2. The modification of the reptilian scales into feathers not only assisted with flight control on the wings, but also was an important adjunct to:
3. The development of homoiothermy.

The Jurassic fossils, so well preserved in the fine-grained limestone of Bavaria, have been named *Archaeopteryx*. Were it not for the feathers preserved *in situ*, the skeleton might well have been classified as that of a reptile, as the skull bones in particular closely resemble those of an archosaur. Even the "beak" possessed well-developed teeth. The fact that the body, wings and tail bore feathers, however, necessitates the classification of *Archaeopteryx* as a bird, albeit a very early one, and suggests that it was warm-blooded.

Nothing more is known of the evolution of birds until the Cretaceous, and by then a considerable radiation had obviously occurred. This is well illustrated by two well-known genera from the chalk of Kansas. Both of these were toothed, and both were sea-birds, but one genus, *Ichthyornis*, must have been a strong flyer, whereas the other, *Hesperornis*, had secondarily already lost the power of flight, and was adapted for swimming and diving.

By Caenozoic times, most of the modern orders of birds had appeared, including some spectacular members of what is considered to be the most primitive living group of birds, the Ratites. Members of this order include the flightless ostriches, rheas and emus, and the extinct moas and elephant birds.

Further Reading

Colbert, E. H. (1951) *The Dinosaur Book*. New York.
Colbert, E. H. (1955) *Evolution of the Vertebrates*. London.
Goodrich, E. S. (1930) *Studies on the Structure and Development of Vertebrates*. London.
Heilmann, G. (1926) *The Origin of Birds*. New York.
Pycraft, W. P. (1910) *A History of Birds*. London.
Romer, A. S. (1955) *The Vertebrate Body*. London.
Romer, A. S. (1955) *Vertebrate Palaeontology*. Chicago.
Romer, A. S. (1956) *Osteology of the Reptiles*. Chicago.
Swinton, W. E. (1958) *Fossil Amphibians and Reptiles*. British Museum, London.
Swinton, W. E. (1958) *Fossil Birds*. British Museum, London.
Young, J. Z. (1950) *The Life of Vertebrates*. Oxford University Press.

8
mammalian radiation and the origin of man

What is a Mammal?

The Mammalia, which we would probably rightly consider to be the dominant animal class today, is a fairly homogenous group anatomically and physiologically, although adaptive radiation within the class has produced an enormous variety of forms. The main features associated with mammalian organisation may be summarised as follows:

1. The presence of a four-chambered heart.
2. The survival of only the left systemic arch.
3. Homoiothermy.
4. Methods of regulating temperature, such as hair and sweat glands.
5. The presence of a muscular diaphragm to aid breathing.
6. Lower jaw consisting only of dentary.
7. Three ear ossicles present.
8. Mammary glands present in female.
9. Brains proportionately larger than in any other group of animals.
10. Allantoic placenta usually present.
11. Specialised dentition, limited to two successive sets of teeth.

These features listed are all fairly obvious, mainly anatomical points, but one could make another list including more detailed physiological and behavioural points. Even of the ten anatomical features listed, however, it is easy to see that very few of them can be applied critically to fossils, in order to distinguish between reptilian and mammalian tetrapods. Points 6, 7, 11, and possibly 9, will feature in palaeontological data, and it is on these that taxonomic decisions will have to be based. In fact, the agreed criterion, which designates a tetrapod as being a reptile or a mammal, is the structure of the lower jaw and its articulation with the skull. If the articular bone is still a component of the lower jaw, and this articulates with the quadrate bone of the skull, then the animal is said to be a reptile; whereas if the articular bone has become modified to form one of the ear ossicles, and the jaw hinges on a joint between the dentary and the squamosal bone of the skull, the animal is a mammal. Obviously this entirely ignores (of necessity) any reference to the structure of the circulatory system or the method of producing young, but as we have seen before, the palaeontologists' concept of taxonomic units is bound to differ from that of a biologist concerned with existing organisms, every aspect of which can be studied.

Origin of Mammals

The mammalian skull is most easily derivable from the synapsid reptilian condition, and as we shall see, there are certain advanced members of this group which are mammal-like in other respects. As far as the class itself is concerned, however, the first undoubted mammalian remains are found in the early Jurassic. These remains consist of skull and jaw fragments of small size, and indeed the largest Jurassic mammal of which remains have been found was about the size of a cat. This small size is undoubtedly correlated with the fact that the reptiles were dominant at this stage, and these primitive mammals must have kept out of the way of the dinosaurs, whom they were later to succeed as the dominant vertebrate group.

Mammalian Ancestors amongst the Synapsid Reptiles

We have already seen how the Permian pelycosaurs evolved into the more advanced Triassic therapsids. The basic stock of this latter group was the suborder Dinocephalia, and this consisted of awkward, massive reptiles, some of which were herbivores, and some of which were carnivores. Radiation occurred within this group according to mode of nutrition, and the herbivores gave rise to the Jurassic and Triassic dicynodonts, which showed some mammalian features such as a rudimentary secondary palate; whilst the carnivorous forms evolved into the theriodonts. These latter reptiles were much closer to the mammalian line of descent, since they developed differentiated teeth, a secondary palate, and the dentary bone was becoming the dominant component of the lower jaw.

The Gorgonopsia, represented by the Permian *Gorgonops*, were the most primitive theriodonts, possessing differentiated teeth, but with only the first traces of a secondary palate. The pineal opening was still present, and the arrangement of lower jaw bones can be seen in the accompanying diagram. From the gorgonopsid stock, two other groups radiated, both of which were still very mammal-like in many respects.

The Therocephalia, and their Triassic successors, the Bauriomorpha, represented by *Lycosuchus* and *Bauria*, formed one of these evolutionary lines; and the cynodonts, represented by *Cynognathus*, the other. *Cynognathus* is a particularly well-known fossil. It was a dog-sized creature, and must have been dog-like in many other respects as well. Although mammalian in many skeletal features, it cannot be regarded as a mammal, as its lower jaw articulation was still by means of a much reduced articular bone.

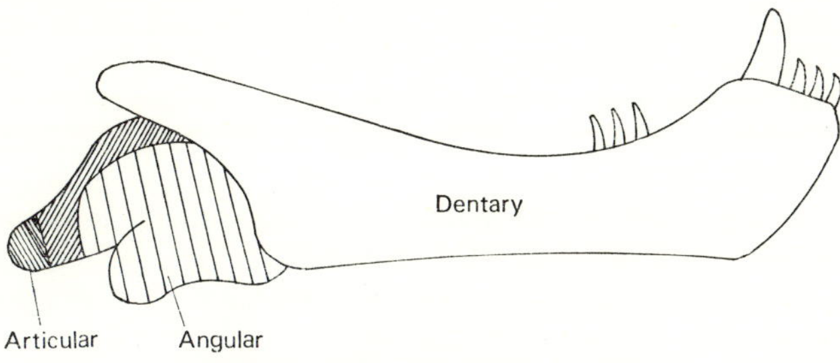

Fig 8.1. Gorgonopsid lower jaw

A fifth group of theriodonts is also known, and these are completely intermediate between reptiles and mammals in all skeletal respects. One of these ictidosaurs, as they are known, is even intermediate with respect to the jaw articulation. *Diarthrognathus*, as its name implies, has both an articular-quadrate and a dentary-squamosal articulation

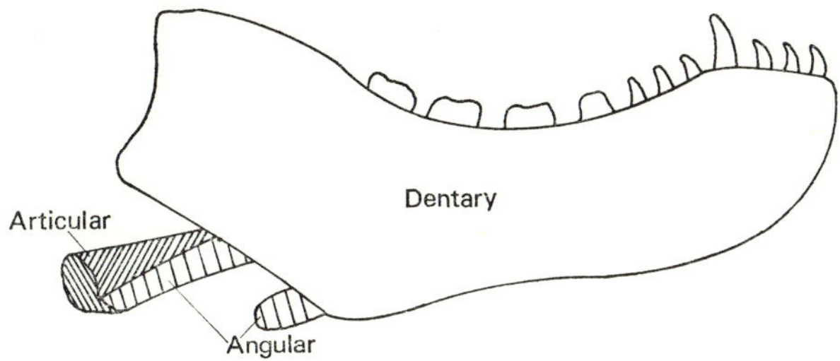

Fig 8.2. Cynodont lower jaw

between the lower jaw and the skull. The affinities of this Triassic group, however, are controversial, as some authorities seek to derive them from the therocephalians, and others from the cynodonts. It is remarkable and significant, how many groups of synapsid reptiles were evolving side by side towards the mammalian condition, and it is quite probable that the first mammals were of polyphyletic origin. Some therocephalians may have crossed the border line to the mammalian state quite independently of cynodonts and ictidosaurs.

Palates, Fur and Ears

We have seen that the development of the secondary palate is one of the main mammalian characters. Why should this be so, especially as we have already observed that a similar structure developed independently in crocodiles? The reason is that the presence of this structure enables breathing to be carried out independently of and simultaneously with feeding. This is necessary in animals with high metabolic rates such as mammals and birds, where breathing cannot efficiently be spasmodic. The necessity for the separation of the two processes in crocodiles is due to the fact that the mouth is frequently operative under water.

Were any of the advanced therocephalians, cynodonts or ictidosaurs warm-blooded, hairy, and with muscular diaphragms? It is impossible to tell from the existing fossil evidence, but if they had high metabolic rates, as is suggested by the palate, then they must have had some temperature regulating features. The arrangement of the posterior thoracic ribs in some ictidosaurs suggests that they may have supported a diaphragm.

The story of the evolution of the mammalian ear is one of the most fascinating cases of bony elements changing their function. The fish spiracle persists in tetrapods as the eustachian tube internally and the external auditory meatus externally, with the tympanum separating the two regions. In lower tetrapods, the tympanum transmits vibrations to the cochlear region by means of a single bone, the columella or stapes, which is derived from part of the fish hyomandibula. In the mammal, however, when the dentary takes over as the lower jaw component articulating with the skull, the articular and quadrate bones being freed from their previous function, become detached from the lower jaw and skull respectively. They enter into the auditory system as the two extra ear ossicles. The incus is derived from the quadrate, and the malleus from the articular. (Mnemonic tip: I.Q. and M.A. are easy pairs of letters to remember.) The only other original lower jaw component remaining is the angular, which becomes the mammalian tympanic bone, supporting the eardrum.

The Earliest Mammals

The relatively few Mesozoic mammals that are known, mainly from the Jurassic sediments of Europe, have been grouped into four orders, chiefly on dental features. These are the Triconodonta, the Symmetrodonta, the Multituberculata, and the Pantotheria.

The first two of these orders consisted of small, rather undistinguished carnivores, and neither of these groups survived long after the close of the Jurassic.

The multituberculates, on the other hand, were rather specialised and successful mammals. Although still small, they survived well into

Caenozoic times, and were certainly the first herbivorous mammals, being remarkably rodent-like in their dentition. Thus they possessed in each jaw a pair of enlarged incisors, behind which there was a diastema (no canines) followed by molars with longitudinal rows of cusps, suitable for chewing or grinding vegetation.

The fourth order of Jurassic mammals is probably the most important of all, as it is fairly certain that these pantotheres gave rise to the marsupial and placental stocks. It must be admitted, however, that this supposition is based almost entirely on the type and number of teeth possessed by these mammals, for very few remains other than fragments of jaws and teeth are known.

Mammalian Radiations

After the initial radiation of the small Jurassic mammals, and the origin of the marsupials and placentals during the Cretaceous, the "big bang" of mammalian radiation came in the post-Cretaceous Caenozoic. The reason for this sudden evolutionary burst was undoubtedly correlated with the remarkable and somewhat inexplicable extinction of all the dominant reptilian forms at the end of the Cretaceous. This wholesale dying out left numerous ecological niches unoccupied, and the mammals were at last able to emerge from their modest position, and become the new dominant terrestrial class. Although there were climatic and geological changes at the close of the Mesozoic, these were not catastrophic, and it remains a mystery why all the ichthyosaurs, plesiosaurs, dinosaurs and pterosaurs should suddenly have died out; but in doing so, they left the way clear for the mammals and eventually man.

During Cretaceous times, as has been mentioned, the earliest marsupials and placentals evolved side by side, and apparently equally successfully. At the close of this period, however, and before the major placental radiation, the minor geological changes that occurred separated Australia from the Asian continent, and also severed the link between North and South America. In Australia, the marsupials became the most successful animal form, radiating into all the niches available to them. A similar course of events happened in South America, except that the marsupials and the placentals maintained their balance to become the co-dominant forms. Elsewhere in the world, however, the placentals became the dominant partner, and their efficiency and specialisation were such that when, at the close of the Tertiary, North and South America became joined once again, the invading placentals from the north exterminated both the southern marsupials and placentals, as a result of competition. Australia has of course remained geographically isolated, but the superior adaptive features of the placentals are shown here equally well, in the cases where man has introduced placentals to the continent. The introduced forms have proved much more successful in their particular ecological niches than their marsupial predecessors. Thus the rabbit has become a major pest in Australia, and the placental dog is ousting the marsupial Tasmanian wolf.

Only very few marsupials have survived outside Australia in direct competition with placentals. The Opossum of the southern states of North America is probably the best known example, and is also noteworthy in that it has remained virtually unchanged since the Cretaceous, being very similar structurally to the original primitive marsupial stock.

The third main group of mammals, the monotremes, are worthy of separate consideration.

Monotremata

There are three living genera of monotremes; the Duck-Billed Platypus (*Ornithorhynchus*) of Australia, and the Spiny Ant-eaters (*Tachyglossus* and *Echidna*) from Australia and New Guinea respectively. These animals are mammalian in so far as they possess hair, a four-chambered heart, a diaphragm, dentary and three ear ossicles, and mammary glands of a primitive type. They also, however, lay eggs, lack external ears, have their rectum and urino-genital system opening into a cloaca, and possess several reptilian skeletal features. In addition to this confusion of characters, they are unknown in the fossil record before the Pleistocene, so tracing their ancestry becomes very difficult. Because the females have brood pouches similar to those of the marsupials, it has been suggested that they were an early offshoot of the marsupial line of evolution, but skeletal features rather preclude this, and it is more likely that they represent a separate line of evolution from the theriodonts. If this is so, our earlier theory as to the polyphyletic origin of mammals is given strength.

Placental Radiation

The Caenozoic is really the era of the placental mammals. The variety of these is so great that it is impossible to consider even all the orders in any detail, but we shall attempt to outline the main radiations that took place. The fossil record of these recent developments is tantalising. In some cases (for example the horse) the story is so well "documented" palaeontologically that evolutionary pedigrees can be traced in some detail over millions of years. In other cases, however, there is an almost total lack of fossil evidence, and a whole new order may suddenly appear with absolutely no indication of its derivation.

For convenience, we shall adopt Simpson's scheme of classification, which groups mammalian orders into four cohorts: the Unguiculata, the Glires, the Mutica, and the Ferungulata.

The Unguiculata

These are mammals possessing nails or claws on their phalanges, and include both the most primitive and the most advanced mammalian orders.

Order Insectivora

This, the most primitive order of mammals, includes all the earliest Cretaceous mammals. Insectivores are all small, with a primitive brain, and the basic mammalian dentition. The order includes tenrecs, shrews, moles, hedgehogs, and numerous extinct genera. From the insectivore stock, all the other placental mammalian orders evolved.

Order Chiroptera

The bats are unique in being the only fully aerial group of mammals. Probably because of this habit, however, their fossil ancestors are very few, and the earliest bat known, from the Eocene, was already fully developed and very similar to modern forms. The early evolutionary history of the group is thus unknown, although it must have developed from the insectivore stock remarkably quickly. There are two distinct suborders of living bats. The Megachiroptera, or Old World fruit-eating bats, which have simple blunt teeth; and the Microchiroptera, which are small, insect-eating bats with sharp cusped teeth, and an acute sense of hearing. This latter group includes the well-known vampire bats, of slightly aberrant feeding habits.

Order Dermoptera

The only living representative of this order is *Galeopithecus*. This is a West Indian mammal which is capable of gliding distances of up to 70 metres by means of folds of skin stretched between fore- and hind-limbs and the tail. Fossil ancestors are known intermediate between the early insectivores and the modern "Flying Lemurs", a name, incidentally, as inaccurate as that of the Holy Roman Empire.

Order Edentata

These unguiculates from America include the armadillo, descended from *Glyptodon*-like ancestors, and the tree-sloths and giant ant-eater, smaller descendents of the elephant-sized Pleistocene ground-sloths. *Manis*, the scaly ant-eaters or pangolins are also closely related to these other living forms.

Order Primates

This group will be considered in more detail later, as it is the order within which man has evolved.

The Glires

This cohort comprises the rodents and rabbits. Originally classified together, it is now realised that the resemblances between these two orders are superficial, and probably the result of convergent evolution.

Order Rodentia

Rodents have primitive brains and are particularly characterised by having two large pairs of chisel-shaped incisors which grow continuously as they are worn down. This enables these herbivores to gnaw very efficiently, as is particularly demonstrated by the beaver. Other rodents are squirrels, chipmunks, marmots, rats, mice, and porcupines.

Order Lagomorpha

The rabbits and hares, although superficially similar to rodents, have four pairs of gnawing incisors, as contrasted with the rodents two pairs, as well as other skeletal differences. The affinities are due to the two groups having adopted the same mode of life, but their evolutionary origins were certainly very different. Although fossil ancestors of the rodents are very scanty, the primitive skull and brain indicate their descent from a very early branch off the insectivore stock. The lagomorphs, on the other hand, are probably derived from a primitive order of hoofed mammals, the condylarths.

Cohort Mutica

This cohort is synonymous with the order Cetacea.

Order Cetacea

The whales, porpoises and dolphins form a fascinating group, both on account of their re-adaptations to the marine environment, and because of the sheer size of some of the species. Thus the well-known blue whale dwarfs even the largest dinosaurs in being the largest animal ever to exist, with a length of over 30 metres, and weight of over 150,000 kg. The cetaceans are also of great interest in having a very high level of intelligence, second only probably to that of the higher primates.

In re-adopting the aquatic existence, the whales have become very un-mammalian in appearance. Apart from the necessary streamlining of the body, which results in loss of differentiation into head, trunk and limbs, the external hair has been lost, its heat-controlling function being taken over by a subcutaneous layer of blubber. The posterior limbs have been completely reduced, whilst the anterior appendages have become modified to flippers. In addition to these reductions and modifications, two entirely new structures arise; a dorsal fin, and a horizontal tail fluke. These two structures are remarkably analagous to the similar structures found in fish and in the ichthyosaurs.

Many physiological adaptations are also present, especially concerned with the respiratory system, enabling cetaceans to remain submerged for periods of time exceeding an hour.

Of the evolutionary history of whales, nothing is known. True whales, very similar to modern forms, appear in the fossil record in early Eocene sediments, but before this time there is no trace of the order. It must have arisen in the late Cretaceous burst of mammalian evolutionary radiation.

Cohort Ferungulata

This group, which contains an apparently heterogenous collection of mammals, is the largest of the four cohorts. It is comprised of all the carnivores and all the hoofed mammals.

Order Carnivora

The carnivores are chiefly characterised by their dentition, with nipping incisors, tearing or stabbing canines, and chewing or shearing pre-molars and molars. They have strong jaws, tend to be relatively small and active, and have keenly developed senses. All these features are very necessary if carnivores are to be successful, for unlike herbivores, their food will not stay still to be eaten.

The earliest carnivores to be found in the fossil record are the creodonts of early Eocene times. These were primitive weasel-like animals, which were virtually extinct by the close of the Eocene. They gave rise, however, to the far more successful group of carnivores, the Fissipedia, which today includes all the terrestrial carnivores, ranging through dogs, bears, pandas, cats, weasels, badgers and hyenas. The name carnivore is used here in a strictly taxonomic sense, for some of this group are omnivorous (e.g. bears) and some purely herbivorous (e.g. pandas). The radiation of the fissipedes produced such extreme forms as the sabre-toothed tiger, and the giant cave-bears of Pleistocene times.

At some time during the Eocene, a group of carnivores returned to the sea, paralleling the action of the cetaceans. The Pinnepedia, however, have never become totally adapted to an aquatic existence, and the sea-lions, walruses and seals all spend a proportion of their time on dry land. They are still carnivores, however, feeding on fish and molluscs.

The Ungulates

This group is a large, complex, and fascinating one, including all those mammals loosely described as having hooves. The category is a somewhat artificial one, as it describes ecological modifications rather than basic affinities, but it is almost certainly true that all the carnivores and all the ungulates derive from a basic Cretaceous ferungulate stock.

The adaptations characterising the hoofed mammals are largely concerned with their herbivorous diet. Thus the teeth have become specialised for biting and grinding, and the alimentary canal has become modified to deal with the plant foods, for example in the possession of a chamber in which symbiotic bacteria can break down cellulose.

Some ungulates have developed enlarged teeth or antlers and horns for defensive purposes, but more commonly the defensive mechanisms developed were those concerned with running away. Such developments included the elongation of the limbs and feet, and the tendency to walk on tiptoe.

To describe the evolutionary history of all the ungulates would be a colossal task, but fortunately, one well-known example illustrates well some of the tendencies exhibited by the majority of hoofed mammals.

Evolution of the Horse

By a series of fortunate circumstances the fossil history of the modern horse is very well known from a set of consecutive sedimentary deposits in North America.

The earliest horse ancestor known from early Eocene times is *Eohippus* (or more correctly but less familiarly, *Hyracotherium*). It was comparatively small, about the size of a modern fox, and was adapted for fast running on slender limbs each with three functional toes. Each toe was capped with a small hoof. The teeth were relatively unspecialised and presumably adapted for browsing. From *Eohippus* to the modern genus, *Equus*, a series of trends are apparent which were probably connected with the gradually changing environment, from lush forest to open dry grassland. This would have necessitated changes in locomotion and feeding habit for any organism to survive as a reasonably sized population. The trends apparent in the line of horse evolution are as follows:

1. Increase in size.
2. Increase in size and complexity of the brain.
3. Elongation of legs and feet.
4. Reduction in number of toes to one middle toe.
5. Increase in complexity of patterns on crown of cheek teeth.
6. Increase in height of crowns of cheek teeth.
7. Consequent increase in length and depth of the face to accommodate the larger teeth.

The process of change was gradual through the series: *Eohippus—Orohippus—Epihippus—Mesohippus—Miohippus*. The last named Oligocene genus was about the size of a sheep, and was probably capable of maintaining a good speed over hard ground. After this somewhat orthogenic sequence, however, the variety of available habitats increased, and a radiation of horse forms occurred, some of which spread to most of the continents of the world. The progress to the present-day horse was continued through *Parahippus* and *Merychippus* in the Miocene, and *Pliohippus* in the Pliocene.

More recently, the evolution of the horse has been somewhat controlled by man's selective breeding, following his initial successful domestication of the animal several thousand years ago.

A Review of the Primates

There is little doubt that the primate group evolved comparatively early in the history of mammals. Thus although the fossil record is tantalisingly scarce, true primates are found in the early Eocene, almost sixty million years ago. The features which distinguish Primates from their insectivore ancestors and their Unguiculate cousins are the forerunners of those very characters that have given man his present-day superiority. All these features were originally modifications for an arboreal life, but they have continued to serve man well since his descent to a more terrestrial existence.

Quick reactions, and a consequent increase in the size of the cerebral hemispheres to deal with the computations necessary for the split-second timing required for movement along and between branches of trees, were two adaptations of the central nervous system. A lessened olfactory sense and an increased visual ability coupled with binocular vision, were the advantageous modifications of the sense organs. And the great mobility of the digits which assisted in the gripping of branches was of immense use later for tool manipulation. Family life, too, became more marked in the Primate stock, together with a reduction in the number of offspring produced.

Living Primates can be considered as belonging to one of two groups:

1. *The Prosimia*

This suborder includes all the lower Primates such as the tree shrews, the lemurs, the lorises, the aye-aye of Madagascar, and the tarsiers. Fossil remains of the ancestors of this group are known from the Eocene, and suggest that the tree-shrews and tarsoids may not have been far removed from the evolutionary line giving rise to the other main Primate group:

2. *The Anthropoidea*

These higher Primates can be further subdivided into three superfamilies.

(*a*) The Ceboidea. The New World monkeys are the most primitive anthropoids, and are characterised by having three premolar teeth and frequently a prehensile tail. These are the marmosets, capuchins, and spider monkeys, and their fossil history is almost non-existent.

(*b*) The Cercopithecoidea, on the other hand, are very well documented palaeontologically. These are the Old World monkeys; baboons, rhesus monkeys and colobines. More advanced and intelligent than the Ceboidea, they have two premolars generally, small ears, and a tail which is never prehensile. The fossil record can be traced from *Parapithecus* of the early Egyptian Oligocene through to the present day.

(*c*) The Hominoidea. This last superfamily contains the families Pongidae and Hominidae. The Pongidae include the four living great apes; the gibbons, chimpanzees, orang-utans, and gorillas, whilst the Hominidae includes only one living species, *Homo sapiens*.

The Fossil Ancestry of Man

Finding and identifying fossil ancestors of man has always been one of palaeontology's "glamour" features. The public interest which this arouses has been alternately fanned and extinguished by discoveries, controversies and the exposure of hoaxes, such as Piltdown Man; but the investigation continues to attract many colourful and able researchers, who work unremittingly to find one more "missing link". Every year brings some new discovery or reassessment of fossils found earlier, so all we can do is to give the most likely and widely accepted views as to the nature of certain ancestral fragments of man.

The history of the various discoveries and arguments is a fascinating study in itself, but there is no room here to touch on this aspect of the topic.

The earliest known fossil showing Hominoid features is a lower jaw bone, between two and three inches long, known as *Propliopithecus*. This was found in the early Oligocene deposits of Egypt, and is over thirty million years old. For many years this was considered to be an ancestral gibbon, but reconsideration of the teeth show that it has both Pongid and Hominid features, and it may, therefore, be close to the point at which the Hominid stock diverged into men and apes. More recently, but in the same deposits, remains have been found of another creature which has been named *Aegyptopithecus*. This was a contemporary of *Propliopithecus*, and yet it showed definite affinities to the Pongid stock, suggesting that the earlier discovered fossil might be more directly in the Hominid line than previously supposed.

The next relevant fossils are found in the Miocene. In the middle of the last century, a fossil was discovered in France which was given the name *Dryopithecus*. Since then numerous dryopithicines have been discovered from sediments all over the world, and these have been split into nearly thirty genera and fifty species. The "lumpers" have been at work recently, however, and have reduced these numbers considerably. Another notable fossil from this era was *Proconsul*, which is now also considered to be a dryopithicine, although originally it was presumed to be on the direct line of descent of man.

All these dryopithicines, however, are definitely of Pongid stock, having evolved from *Aegyptopithecus* or a near relative, and having given rise to the present day great apes. Indeed, certain late Miocene dryopithecines can be placed fairly accurately as the ancestors of the gorilla, the orang-utan, and the chimpanzee. The gibbons had almost certainly diverged earlier, being derived from forms very similar to the mid-Pliocene *Pliopithecus* and *Limnopithecus*.

A late contemporary of the dryopothecines was *Ramapithecus* of Africa and India. This is a very interesting fossil, as it was almost certainly a true Hominid, and thus the earliest direct ancestor of man known, being 14 million years old.

At this point there is a comparatively long gap in the fossil record, for no further stages in the evolution of man are known until early Pleistocene times, rather less than a million years ago. *Australopithecus* is the first of these later forms to appear, some 600,000 years ago; the first fossils having been discovered in 1924 in Africa by Dart. The name *Australopithecus*, which means "southern ape" is probably rather misleading, for it is now thought that this animal was very close to the Hominid stock. It walked erect, had a simple pebble tool culture, and very probably had the power of speech. Its average brain size was 600 cubic centimetres, compared with modern man's 1,400 cubic centimetres. Dr Louis Leakey has recently discovered two fossils of Hominids, which he has christened *Zinjanthropus* and *Homo habilis* respectively. Most anthropologists would prefer to regard these forms as australopithecines, but they are none the less interesting, as they would seem to push back the australopithecine ancestry into the Pliocene, between one and two million years ago.

Pithecanthropoids

In 1890, a young Dutch anthropologist named Dubois discovered some teeth and a skull-cap in some mid-Pleistocene sediments in Java. More bones were discovered later which showed the human affinities of the creature. The fossil was named *Pithecanthropus erectus* or the upright "ape-man", nowadays often known simply as Java Man.

Thirty years later, some further Hominid remains were found in Pleistocene infillings of fissures in earlier Ordovician limestone. This discovery was made by Black at Choukoutien near Pekin. Black named this fossil *Sinanthropus pekinensis*, but recent work has shown that it is very similar indeed to *Pithecanthropus* and has accordingly been moved to that genus, maintaining its original specific name. This is Pekin Man.

Pithecanthropoids have been found also in Algeria, Morocco and Heidelberg, showing that the range of this Hominid was not restricted to Eastern Asia.

Structurally, *Pithecanthropus* was much closer to modern man than were the australopithecines. Remains have been found associated with

Pekin Man, however, which suggest that the affinities were cultural as well as anatomical. Tools shaped from quartz pebbles and bones, were found in association with Pekin Man, as were ashes and charred bones, showing that not only did *Pithecanthropus* have fire, but that he probably used it for cooking. Some of the bones present in the remains show that Pekin Man was a hunter, both of large animals such as deer, and of his own kind. Split-open skulls testify to this last belief.

Finally, one of the most interesting points is that seven Choukoutien skeletons were found together underneath a scattered layer of red ochre, a suggestion that this might have been the start of some kind of ritual burial of the dead. Unfortunately, very few artifacts have been found associated with Pithecanthropoid remains outside China. All the evidence is now lost, as an American train carrying all the known remains of Pekin Man to safety "disappeared" outside Pekin in 1941.

Neanderthal Man

Various remains have been found during the last 200 years in Europe, Africa and Asia of a Primate so man-like in most features that they have been placed in the same genus *Homo*. The specific name, however, is derived from a valley in Germany where the first recognised remains were found, the Neanderthal.

Homo neanderthalensis lived between 150,000 and 40,000 years ago, and although off the main line of evolution of *Homo sapiens*, was very advanced in many respects. He walked completely upright and grew to a maximum height of 1½ metres. He made relatively advanced flake tools, very comparable to those used today by the Australian aborigines, and used some of these as weapons, with which he hunted down the largest animals of his time, such as the woolly rhinoceros, the mammoth, and the cave-bear. He almost certainly used animal pelts as clothing, and practised several forms of ritual behaviour, such as burying the dead facing in a specific direction (west) and collecting cave-bear skulls in primitive patterns.

Neanderthal Man was a very variable species, and forms from different parts of the world differ greatly in their skull structure. It is thus possible that there were distinct geographical races of *Homo neanderthalensis*, or there may even have been interbreeding with the earliest members of the species *Homo sapiens*.

Cro-Magnon Man

About 40,000 years ago, a new Hominid appeared on the European scene. Popularly known as Cro-Magnon Man, after the French rock-shelter, where the first specimens were discovered, he is more importantly called *Homo sapiens*. Here then, is the first appearance chronologically of a form of creature almost indistinguishable from modern man. A high forehead, a "flat" face and a large cranial capacity were the criteria distinguishing Cro-Magnon Man from his Neanderthal contemporaries, and although physically weaker, he had gained complete superiority within 10,000 years, when Neanderthal Man became extinct. Several varieties of Cro-Magnon Man are known. Two skeletons from Grimaldi, in southern France, have almost negroid features, whilst a skull from near Perigeux is reminiscent of present-day Eskimos. It is still uncertain, however, whether these represented distinct races of Cro-Magnon Man, or whether they are merely indicative of great individual variability within the species. Where these first members of our own species originated from is still uncertain, but when they reached Europe, they remained and prospered, after they had disposed of their rivals, Neanderthal Man.

Whether Cro-Magnon Man was universally distributed throughout the world, and eventually gave rise to all the present-day human races is not clear. It is equally possible that he was the forerunner of only the Caucasian stock. Nor is it clear when the original Hominid stock diverged to give rise to lines resulting in Neanderthal and Cro-Magnon Man, but it is likely that this branching occurred comparatively early on in the Quaternary period.

Further Reading

Broom, R. (1932) *The Mammal-like Reptiles of South Africa*. London.
Carrington, R. (1963) *A Million Years of Man*. London.
Clark, W. E. Le Gros. (1959) *The Antecedents of Man*. Edinburgh.
Clark, W. E. Le Gros. (1960) *History of the Primates*. British Museum, London.
Colbert, E. H. (1955) *Evolution of the Vertebrates*. London.
Goodrich, E. S. (1930) *Studies on the Structure and Development of Vertebrates*. London.
Howells, W. (1960) *Mankind in the Making*. London.
Romer, A. S. (1955) *The Vertebrate Body*. London.
Young, J. Z. (1950) *The Life of Vertebrates*. Oxford University Press.
Young, J. Z. (1957) *The Life of Mammals*. Oxford University Press.

9
the early evolution of plants

The study of the evolution of plants is based on the same types of evidence and is subject to the same limitations in general as is the evolutionary study of animals. Further complications, however, are introduced by the following facts:

1. Higher plants cannot move, and therefore cannot stand much chance of being fossilised if they grow in regions not favourable to the processes involved in preservation. An animal, in contrast, may easily fall into a stream or mudpool which will preserve its remains, although this accident may occur far from the animal's normal habitat.
2. Plants tend to detach component parts of themselves much more readily than do animals. Thus a fossilised leaf and a fossilised pollen grain may be assigned to separate species, because of lack of organic connection between them, although in fact derived from the same plant.
3. Plants in general do not have hard parts comparable to the skeleton of animals, and therefore are likely to be preserved in far fewer situations.
4. The range of structural form exhibited by plants is very small compared with that occurring in the animal kingdom.

The last mentioned point is an advantage as well as a disadvantage, for it does mean that there are no major groups of plants whose affinities are completely unknown, although interrelationships within the groups are sometimes unclear.

We shall discuss plant evolution first by describing the major groups of plants and their occurrence in the fossil record, and then by considering the possible course of structural evolution.

The plant kingdom is subdivided into a number of groups, each of which represents a distinct structural division. The latter term is often applied to these groups, though there is no valid reason why they should not be designated phyla, to parallel the terminology used in the animal kingdom.

Thallophyta

The simplest groups of plants are found in this division. They are characterised by a plant body or thallus which is not differentiated into stems, roots and leaves. Different authorities place different plant groups in this division. Thus the algae, fungi, lichens, bacteria, myxomycetes and viruses are all included by some. We have discussed in Chapter 2 the possible origins of heterotrophs and autotrophs, and their relationships. We have also considered briefly the position of viruses, and from knowledge of their biology it is clear that they have played no direct part in the evolutionary history of plants. The fungi shall be omitted from this division and considered separately later. This also excludes the symbiotic lichens and the myxomycetes, so my division Thallophyta only includes the bacteria and the algae. Amongst the latter group of organisms, *Euglena* and its relatives having been bandied to and fro by zoologists and botanists are now generally regarded by botanists as being flagellates, and therefore not algae but protistans. Whether the autotrophic protistans are included in the Thallophyta or not is a further moot point.

Algae

In older classifications, the algae are divided into four groups; the blue-green, green, red, and brown algae. This taxonomic sorting depended entirely on the nature of the photosynthetic pigments possessed by the plant, and it is now known that this not only misplaced some true affinities, but also did not take into account other groups of simple plants now included in the algae. Photosynthetic pigments are, after all, merely adaptations to a particular habitat. This is shown most clearly in the zonation found on rocky shores, where green algae appear at the top of the shore, brown algae half way down, and red forms at the lowest limits of plant life. This zonation is correlated with the wavelengths of light able to penetrate the average depths of water covering these plants, and clearly the use of pigments alone as taxonomic criteria is likely to produce an artificial classification. Recent classifications have taken into account not only the nature of the pigments present, but also the main food storage products, the cell wall composition, the presence and nature of flagella, and various other morphological and biochemical criteria. A consideration of all these factors has led to a modern classification in which ten main groups of algae are recognised, all of more or less equal status. Within each of these groups, a range of vegetative and reproductive form is found, from the planktonic unicells to the giant seaweeds such as *Macrocystis*, which can grow to a length of 180 metres.

The blue-green algae differ markedly from all other groups of algae (and higher plants) in several respects. They are unique in having no distinct nucleus, separated from the cytoplasm by membranes, and neither do they possess discrete mitochondria or chloroplasts. They have no sexual means of reproduction, and even their asexual division is not achieved by a normal mitosis. They never possess flagella. Blue-green algae are the first plants to appear in the fossil record, in early Cambrian times, and there is little doubt that the group represents a very primitive stock. There are many remarkable similarities between the blue-green algae and the bacteria, both of a positive and a negative nature, and it is very probable that these two groups of organisms are fairly closely related. What is not at all clear, however, is which of the two types of organism is derived from the other, nor is their relationship with other plants clear, as the blue-green algae possess chlorophyll *a*, in common with all other plants, and this is presumably derived from a common ancestor.

Taking all the other groups of algae, certain evolutionary tendencies can be observed, which are repeated many times.

1. Vegetative Form. The obvious morphological trend is from a unicellular state to a multicellular organisation, and this is particularly clear in the green algae or Chlorophyceae, which are commonly believed to be ancestral to higher plants. The most primitive forms are probably the motile (flagellate) unicells, such as *Chlamydomonas*, and various lines can be derived from these forms resulting in different types of multicellular organisation. One of these lines is the *Pandorina*, *Eudorina*, *Volvox* sequence, resulting in an organised spherical colony, which has fairly obvious mechanical limitations to any further development. Other developments of the motile unicell were probably the unorganised gelatinous colonies such as *Tetraspora*, and the dendroid colonies, e.g. *Prasinocladus*. Further progress in the direction of multicellular development, however, probably came, not from the motile unicells, but from their sedentary descendants. An attached unicellular alga, such as are found in the Chlorococcales, might well have given rise first to filamentous types (e.g. *Spirogyra*), and then subsequently the parenchymatous types such as *Ulva* and *Enteromorpha*.

2. Development of Sexual Reproduction. Another trend seen in the algae is concerned with their reproduction. The simplest type of sexual union found in algae occurs between two flagellate gametes which are morphologically indistinguishable from one another. This is exemplified by several species of the genus *Chlamydomonas*, and because of the essential similarity of the gametes, the process of fusion is known as isogamy. Within the same genus, however, pairs of gametes are known which although morphologically similar, differ physio-

logically and behaviourally, and heterothallism also occurs. *Chlamydomonas Braunii* shows a further advance in being anisogamous, having macrogametes and microgametes, both still flagellate. The final stage in the differentiation of the gametes comes with oogamy, where one of the gametes loses all power of active movement and becomes an egg. Even in oogamy, however, there are several stages, ranging from the complete shedding of the egg in *Fucus*, through the situation in *Laminaria* where the egg is shed from the oogonium but retains an attachment to the plant, to the ultimate situation found in the red algae, where the egg never leaves the oogonium and is fertilised *in situ*. The trend, therefore, is from isogamy to complete oogamy.

3. LIFE-CYCLES. Algal life-cycles are exceedingly complex and at the same time extraordinarily interesting. The study of one or two individual life-cycles can be very misleading and a much wider consideration rewards the student.

The green algae are comparatively straightforward. Thus the life-cycle of *Chlamydomonas* and its relatives consists largely of a haploid organism forming a diploid zygote briefly followed by an immediate meiotic division to give four flagellate zoospores. These then grow to form a normal haploid individual. *Ulva* demonstrates true alternation of generations in that there are two morphologically indistinguishable forms of the plant, one of them haploid, the other diploid. The first form produces gametes which fuse to give a zygote. This then germinates to give the diploid form which produces zoospores meiotically. The zoospores then develop into the haploid *Ulva* thallus. Such an alternation of generations is described as being isomorphic.

In the brown algae further complications may occur. The gametes and zoospores may often be morphologically indistinguishable from one another, but they derive from distinctive organs; the zoospores from unilocular sporangia, and the gametes from plurilocular gametangia. These different origins may be very helpful, for apart from appearing visually similar, the two types of "swarmer" may behave in very confusing ways. For example, haploid gametes may behave as zoospores and germinate to give another haploid thallus; also meiosis may fail to occur in the unilocular sporangia with the result that diploid zoospores are produced which grow to another diploid thallus. Both situations can occur in the same plant (e.g. *Ectocarpus*) in different parts of the world, which only adds to the confusion.

The usual "A"-level type brown alga is very pedestrian in its life-cycle. *Fucus* is diploid and produces gametes meiotically in its gametangia. These then fuse and the zygote grows into another diploid thallus. This life-cycle is curiously similar to the situation occurring in most animals. The most spectacular brown alga seen on British shores is *Laminaria*, the Oar Weed. This is an interesting plant as it does show alternation of generations, but of the heteromorphic type. The huge kelp itself is the diploid sporophyte, and the gametophyte, rarely acknowledged, and even more rarely seen, is a minute filamentous entity. The crinkly oarweed, *Saccorhiza*, is also heteromorphic, but in addition the microscopic gametophyte is dioecious, having separate male and female filaments. In *Cutleria*, another brown alga, the two generations are reversed in their relative sizes. The sporophyte is small, and the gametophyte larger, and presumably because of this unusual situation, the two stages were originally classified as different genera.

Finally in the red algae we find some completely new developments. Admittedly, there is frequently an isomorphic alternation between diploid sporophyte and haploid gametophyte, but in these cases the spores are not motile and are formed as relatively huge tetraspores (groups of four spores, each the result of the meiotic division of a single cell). The advantage of the large spore is that it can provide food for the developing gametophyte for a longer period of time than can the minute zoospore, and therefore allow the red alga to establish itself in positions untenable by other groups of algae. In many red algae, not only the tetraspores are non-motile, but the sperms also lack flagella. If they contact an oogonium, however, they adhere, and the sperm nucleus fertilises the ovum *in situ* on its parent gametophyte. Furthermore, the zygote then develops partly parasitically on the female gametophyte and produces a second sporophyte generation, the carposporophyte. This then produces diploid carpospores, which escape from their sporangium on the gametophyte and germinate elsewhere to produce a diploid tetrasporophyte which can then reinstitute the cycle.

Such is the variety of life-cycles amongst the algae. The situation appears most perplexing, for almost every possible variation of the basic alternation of haploid and diploid generations is apparent in the different algal groups, and each life-cycle is clearly relevant to the mode of life and habitat of its possessor. The general consensus of opinion, however, is that life-cycles in which the vegetative plant is haploid, and in which the diploid stage is represented only by the zygote, are the most primitive. Next, it is thought, isomorphic and heteromorphic life-histories developed, and finally the most advanced state reached is that in which the vegetative plant is diploid, and the haploid stage is represented only by the gametes.

Considering the algae as a whole, it is believed that all groups originated from unicellular flagellate ancestors. Whether or not this origin was mono- or poly-phyletic, the vegetative trend was towards a multicellular attached parenchymatous structure, and this tendency was paralleled in each of the main groups. The fact that no unicellular brown algae survive today is not taken as discounting this theory. At the same time as the vegetative form was increasing in complexity, the reproductive system was evolving towards oogamy and a diploid vegetative thallus.

Bryophyta

The next major division of the plant kingdom to be considered is both smaller numerically and less robust vegetatively than the algae. The bryophytes include the mosses (Musci) and liverworts (Hepaticae), and are characterised by their lack of vascular tissue, and their possession of anchoring rhizoids. The chief feature distinguishing them from all other groups of plants, however, is their life-cycle; an alternation

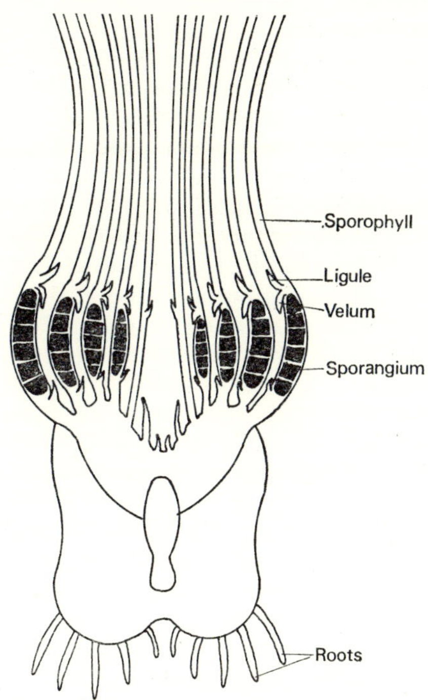

Fig 9.1. *Isoetes* plant (longitudinal section)

of generations in which the diploid sporophyte develops parasitically on the haploid vegetative gametophyte. Fertilisation is effected by a free-swimming male gamete making its way through a film of surface moisture to the archegonium where the ovum is fertilised *in situ*. The necessity for free water has severely limited the potential spread of this group of plants, and they are generally found in the vicinity of free bodies of water or in positions which are frequently damp due to rain or condensation. The method of fertilisation and the lack of vascular tissue have limited not only the distribution but also the size attainable by these plants, although one aquatic species, *Polytrichum commune*, living in New Zealand can grow to 2 metres long.

Although the bryophytes represent a hypothetical intermediate step in the conquest of the land by plants, there is absolutely no evidence of any kind which suggests that they are derived from algae. Nor is there any indication that they gave rise to any groups of higher plants. Furthermore, within the division itself, nothing is known which connects the mosses and liverworts evolutionarily, either in the way of a living intermediate "missing link" form, or in the fossil record. The distinction between the Musci and Hepaticae is very real, however, and includes differences in development of sex organs, in mode of growth and in rhizoid structure; and differences in the development and structure of the sporophyte, and the method of dispersal of the spores. The fossil record indicates that the mosses and liverworts were both separately established by the end of the Palaeozoic, and although many fossil forms have no living descendants, several modern genera are easily recognisable early on in the palaeontological record. The problem of genus definition is itself quite an interesting one in this group of structurally limited plants, and generic differentiation may be based on criteria which would be considered trivial in higher plants.

In the light of the difficulties outlined above, there are considerable problems associated with attempting to discern evolutionary trends within either of the two main groups of bryophytes. The liverworts, however, serve as a good example of how cytogenetic study can throw light on evolutionary relationships, when all other comparative studies seem fruitless. Berrie's work on liverworts showed that three-quarters of the species he investigated possessed a haploid number of 9 chromosomes. Other complements found were $n = 5$ and $n = 10$. It is thus reasonable to speculate that an original basic haploid number of 5 chromosomes became doubled to 10 and later reduced to 9. If this is the case, then the genera of liverworts with these respective chromosome numbers could be placed in a tentative evolutionary sequence. This is a good example of how cytogenetic studies can help to clarify a confused taxonomic evolutionary situation.

We shall not go into further details of relationships within the two main groups of bryophytes, as this would involve the taxonomic differentiation of groups unlikely to be known by the student, and hence be a vast task. Suffice it to say that even the direction of evolution within the main classes, as deduced from the study of vegetative structure, is a matter of some controversy. Earlier workers, such as Cavers, assumed that the evolution of bryophytes was the story of the "invention" of a diploid sporophyte generation which gradually gained the ascendency until it almost attained independence in *Anthoceros*. More recently, however, it has been postulated that the basic Bryophyte life-history is one of an isomorphic alternation of generations which has become unbalanced in both directions. Both these theories place the mosses and liverworts in a position intermediate between the Thallophyte and Pteridophyte phyla, but a more startling modern theory is that in fact the bryophytes are simplified descendants of some of the early pteridophytes, namely the Psilophytales, rather than elaborated green thallophytes.

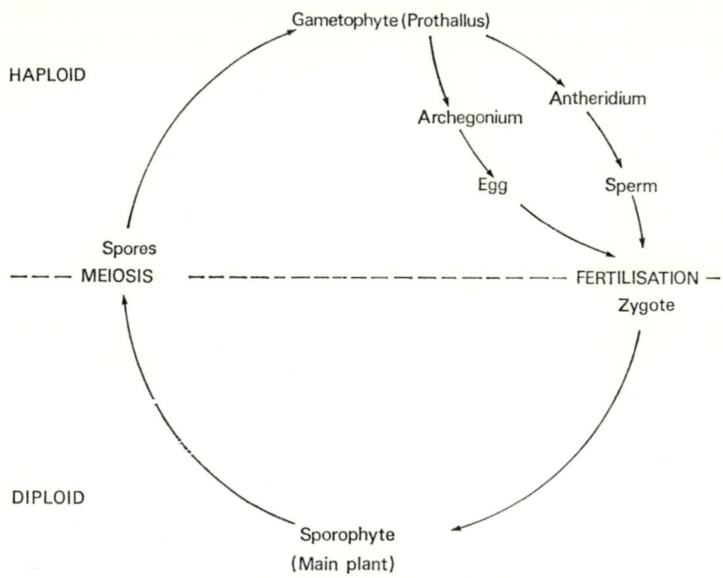

Fig 9.2. Bryophyte life-cycle

The situation concerning bryophytes is open, to say the least, but there is some interesting evidence provided by vestigial structures such as the functionless stomata on the sporophyte of *Sphagnum*, which suggest some tantalising leads without any more corroborative evidence. Perhaps the only hope of shedding any more light on the evolution of the Bryophyta lies in the finding of some earlier fossil remains than are known at present.

Pteridophyta

The members of this next major division of the plant kingdom have life-cycles essentially similar to those of bryophytes, with a regular alternation of haploid gametophyte and diploid sporophyte generations. The crucial and diagnostic difference, however, is that in pteridophytes the sporophyte generation becomes dominant and completely independent of the gametophyte. The sporophyte also becomes larger and considerably more differentiated with internal division of labour producing xylem and phloem, and morphological differentiation resulting in roots, stems and leaves, the latter bearing a cuticle and stomata.

As a result of this new complexity of structure, the sporophyte theoretically can exist in a wide range of terrestrial habitats. Unfortunately, however, the gametophyte often provides a weak link, for it can only germinate, and fertilisation can only occur in environments which are very moist. Some sporophytes overcome this limitation by vegetatively spreading away from their site of initial germination, others solve the problem further back, by having subterranean gametophytes, or gametophytes which are retained within their parent spore

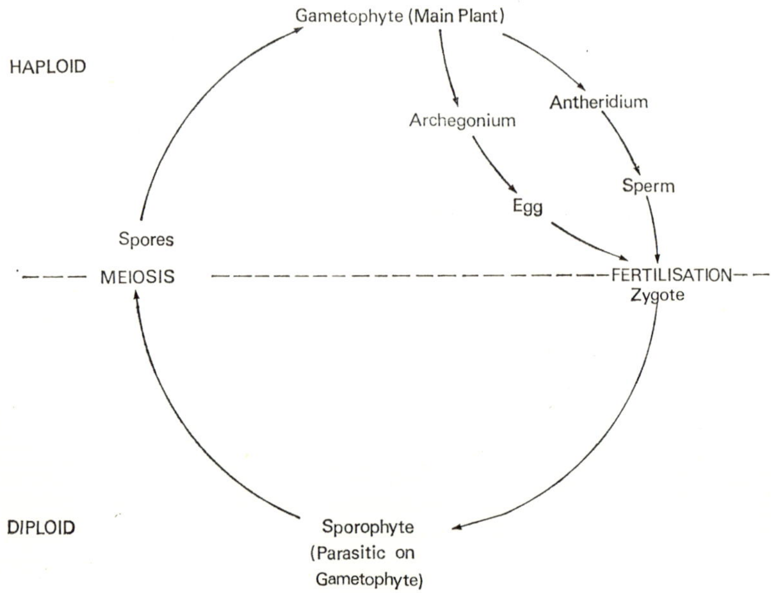

Fig 9.3. Pteridophyte life-cycle

and thus protected from desiccation. In the latter circumstance, heterospory is generally the rule, with megaspores and microspores producing archegoniate and antheridial gametophytes respectively. The significance of this will be discussed later.

The pteridophytes can be conveniently subdivided into five main groups of varying sizes.

Psilophytopsida

Although this group does not contain the earliest pteridophytes known, it does have the distinction of including the first fossil pteridophyte to be discovered, *Psilophyton princeps*. Other genera are known from the Middle Devonian chert of Rhynie in Scotland, and are preserved in such detail that their internal anatomy is known almost as well as that of living plants. *Rhynia*, *Horneophyton*, and *Asteroxylon* are three important genera preserved in this way. This group is the only one of the five major subdivisions of the pteridophytes that is totally extinct. The typical structure of a Psilophytopsid comprised a horizontal, rootless rhizome producing at intervals aerial stems up to 50 centimetres high. These were often naked, though *Asteroxylon* bore many tiny spirally arranged leaves, and bore thick walled sporangia either terminally or laterally. Internally the vascular tissues of the sporophyte

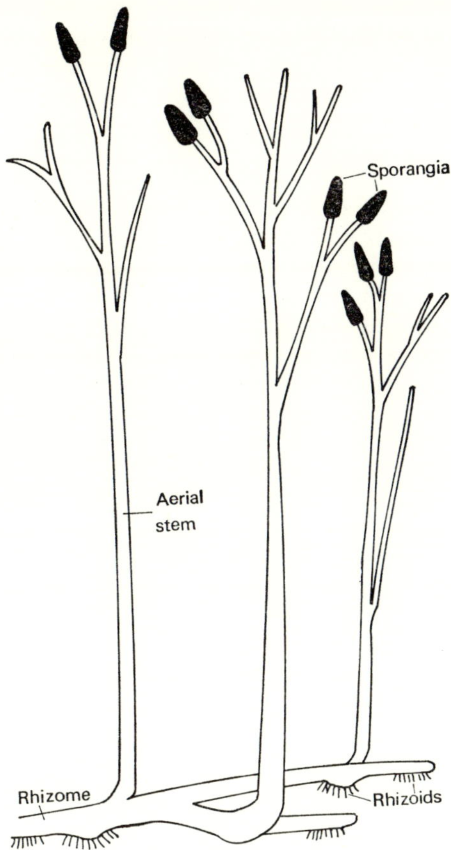

Fig 9.4. ***Rhynia major***

stem were in the form of a solid cylinder, a condition known as protostelic, which is believed to be primitive. The gametophyte is not known, although it is clear that the Psilophytopsids were homosporous.

Psilotopsida

This is a small group of living pteridophytes including only four species, all natives of tropical and sub-tropical regions, and commonly epiphytic. Diagnostically they are very similar to the Psilophytopsida, and are of great interest because they appear to represent a very early stage in the evolution of pteridophytes. Characteristically, there is much controversy over this, and it may be that they represent only the result of extensive reduction. The most interesting feature about them, however, is the remarkable similarity between the gametophyte prothallus and the rhizome of the sporophyte. A gametophyte with internal vascularisation has even been found, which was subsequently discovered to be diploid, and this unusual analogy between sporophyte and gametophyte has been used to support the hypothesis that the original form of the land-plant life-cycle was an isomorphic alternation of generations.

Lycopsida

The oldest known pteridophyte (*Aldanophyton* from the Siberian Cambrian) is a member of this group, as are the living genera *Lycopodium*, *Selaginella*, and *Isoetes*, and the well-known Carboniferous tree fossils *Lepidodendron* and *Sigillaria*. The range of form shown within the group is clearly enormous, but some interesting reproductive developments occur as well. The most important of these are the appearance of heterospory, and the differentiation of spore-bearing leaves (sporophylls) from normal photosynthetic leaves. *Isoetes* shows what is probably the most primitive form of heterospory. The sporophylls, which are photosynthetic as well in this case, are produced in a manner somewhat resembling the growth of bulbs amongst flowering plants. The sporangia form at the base of each leaf, and early in the year contain several hundred megaspores each. Later in the year, the sporangia subsequently produced contain up to a million microspores. Both types of spore are released by the death of the enclosing sporophyll (Fig. 9.1).

Selaginella represents what may have been the next evolutionary step, although there is no suggestion that it is actually descended from such an ancestor. In the 700 or so species within this genus, the leaves born at the tips of the aerial shoots are all sporophylls, and hence this region is called the cone. Many microspores are formed, but on average only one megaspore mother cell matures, so that a single tetrad of four megaspores is formed. The actual numerical range within the genus is from one to twelve or more. This reduction in numbers of both micro- and megaspores is associated with one other remarkable advance. In both types of spore, the gametophyte generation is formed inside the original spore wall by internal divisions of the spore cytoplasm, and in one or two species, the megaspore is never shed from the sporophyte, with the result that the gametophyte and gametes are produced still attached to the parent sporophyte. Fertilisation having been effected, the second generation sporophyte sprouts from the cone of the previous sporophyte generation. The whole of this evolutionary sequence clearly offers tempting clues as to the possible origin of seeds, and more hints are provided by the fossil *Lepidodendron* which was heterosporous, and probably shed its megaspores with the sporophyll still attached. Could this be the precursor of the seed-coat?

Sphenopsida

The fourth main subdivision of the pteridophytes contains several extinct groups including the Carboniferous *Calamites*, which paralleled the tree-like form and development of secondary thickening of *Lepidodendron*. The only living representatives of this group are the members of the genus *Equisetum*. These are the horse-tails, and they show reduction in the structure of their leaves as well as probable reduction from a heterosporous to an apparently homosporous condition. They form an interesting but undoubtedly specialised group.

Pteropsida

This largest of all the pteridophyte groups contains all the living ferns as well as many extinct smaller groups. The range of form and habit is such that it is impossible to define the group precisely, and some authorities would like to include the gymnosperms and angiosperms in the Pteropsida. All degrees of homospory and heterospory are found, and with one exception, no further evolutionary clues of a fundamental nature are offered.

The exception referred to is found in a rather strange little group of aquatic pteropsids. These "water-ferns" are very reduced in structure, but in three genera, which are heterosporous, the sporangia are enclosed in a special structure known as a sporocarp. The wall of this sporocarp shows remarkable anatomical analogies to the seed-coats of flowering plants. Furthermore, in another species, *Azolla filiculoides*, which grows in Britain, the young megasporangium in the sporocarp bears great resemblance to the ovule of flowering plants. These similarities may be coincidental, but it may be that they provide an evolutionary hint.

Further Reading

Bower, F. O. (1908) *The Origin of a Land Flora*. London.

Chapman, V. J. (1962) *The Algae*. London.

Corner, E. J. H. (1964) *The Life of Plants*. London.

Fritsch, F. E. (1935 and 1945) *The Structure and Reproduction of the Algae. Vols. I and II.* Cambridge University Press.

Manton, I. (1950) *Problems of Cytology and Evolution in the Pteridophytes*. Cambridge University Press.

McClean, R. C. and Ivimey-Cook, W. R. (1951 and 1967) *Textbook of Theoretical Botany*. Vols. I and III. London.

Morris, I. (1967) *An Introduction to the Algae*. London.

Sporne, K. R. (1966) *The Morphology of Pteridophytes*. London.

Walton, J. (1953) *An Introduction to the Study of Fossil Plants*. London.

Watson, E. V. (1964) *The Structure and Life of Bryophytes*. London.

Further references to individual papers will be found in all the above general accounts.

10
gymnosperms, angiosperms and fungi

The seed plants, or spermatophytes, represent the ultimate in plant evolution. The green plant has here at last been freed from the limitations imposed on it by its Achilles heel, the free-living gametophyte, with its dependence on copious supplies of free water for growth and fertilisation. The situation found in the bryophytes has been completely reversed, with the haploid generation now being completely parasitic on the diploid sporophyte instead of vice versa. The two great advances shown by the spermatophytes are the development of a pollen tube, which allows internal fertilisation to occur, and the development of the seed as a method of dispersing the protected zygote.

The seed plants are commonly divided into two groups, the gymnosperms and the angiosperms; and these groups yet again illustrate different grades of seed plant development without in any way implying ancestral relationships between living forms.

Gymnosperms

These seed plants are characterised by bearing fully exposed ovules. This is the fundamental distinction between them and the angiosperms, although there are several other less important differences in morphology and anatomy. The gymnosperms themselves can be considered as consisting of three main groups; the Cycadopsida, the Coniferopsida, and the Gnetopsida. The last mentioned category contains three genera of very interesting plants, which have been suggested by some as the collateral descendants of the flowering plant (angiosperm) ancestors because of their structural approach to the higher seed plants. This is unlikely because of the extreme specialisation of the living genera, *Gnetum*, *Welwitschia* and *Ephedra*, but they do provide clues as to the possible derivation of the typical angiosperm characters. The Gnetales, however, as well as being a small group in terms of numbers, is also a group with a very short palaeontological history, as the earliest fossil remnants known are some possible pollen grains from the Eocene and Permian. The other two groups of gymnosperms, in contrast, are large distinct assemblages of plants with very long fossil histories. Thus the Cycadopsida and Coniferopsida appear first in the Devonian, and since then they have remained distinct groups, implying that the gymnosperms may have been derived polyphyletically from the Pteridophyta. The differences between the two main groups are tabulated below.

CYCADOPSIDA	CONIFEROPSIDA
Frond-like leaves.	Needle-, paddle-, or fan-shaped leaves.
Radially symmetrical seeds.	Bilaterally symmetrical seeds.
Soft, sparse, secondary wood.	Dense secondary wood with few small wood-rays.

Living representatives of the two groups are the Cycads only for the Cycadopsida, and the Coniferales, Taxales, and Ginkgoales for the Coniferopsida. Several totally extinct fossil groups are also known, including the Bennettitales and the Cordaitales.

Angiosperms

The flowering plants represent the highest form of seed plants. Here the ovules are protected by being produced inside an ovary, and the sporophylls are borne in definite flowers instead of the gymnosperm and pteridophyte cone arrangement. Because of the enclosed nature of the ovules, the seed always develops inside a fruit, derived from the ovary. The monocotyledons and dicotyledons form the two subdivisions of the angiosperms.

The Seed

As mentioned already, the seed is one of the two major "inventions" of the Spermatophyta. It consists of the fertilised ovule enclosed in a protective coat derived from the integuments, and sometimes containing triploid endosperm. We shall consider the derivation in more detail later, but its importance to the spermatophytes is that it in effect takes the place of the spore in the lower plants as a distributive mechanism. Its advantage over the spore is largely due to the fact that it is multicellular, and thus on germination can produce a relatively large plant from the food-reserves before photosynthetic activity starts.

Alternation of Generations in Spermatophytes

The basic life-cycle as seen in the bryophytes and pteridophytes is continued in to the spermatophytes, although the emphasis has changed. The seed-plant proper is the sporophyte, which is always heterosporous, and which bears its micro- and mega-sporangia on specially modified appendages known as sporophylls. Contrary to widely held belief, however, sporangia are not invariably borne on leaf-like appendages. In some conifers and pteridophytes there is no doubt that sporangia are produced primitively directly on the stem, and it is also doubtful whether all sporophylls are truly homologous with leaves. Whatever their derivation, however, it is convenient to refer

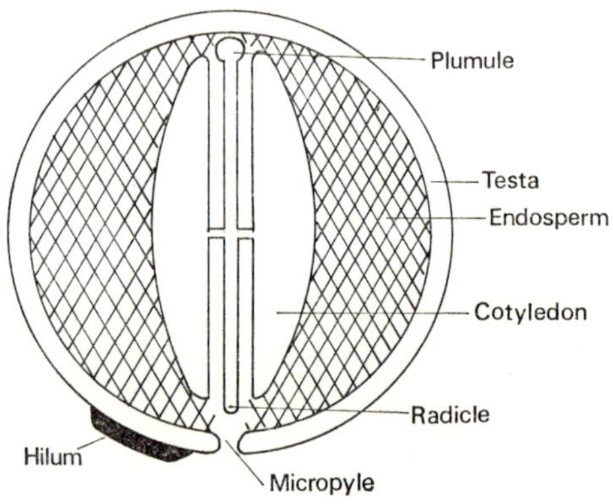

Fig 10.1. The seed

to the structures on which the spermatophyte sporangia are borne, as sporophylls. In most gymnosperms, the micro- and mega-sporophylls are gathered into cones or strobili. The male cones consist of microsporophylls bearing microsporangia, or pollen-sacs. The latter contain at first diploid pollen mother cells which undergo meiotic division to form tetrads of microspores or pollen grains, which are uninucleate at this stage. The female cones are composed of ovules borne exposed on megasporophylls. The ovule is not equivalent to the megasporangium, however, for new elements, the integuments are introduced, which protect the nucellus (or megasporangium proper) to a certain extent. The megaspore (or embryosac) develops inside the nucellus by the megaspore mother cell undergoing a meiotic division followed by the abortion of three of the daughter cells produced. We have thus so far derived the micro- and mega-sporangium and spores in gymnosperms. The next stage in the life-cycle is the germination of the spores to produce haploid gametophytes or prothalli. We have already seen how, in certain pteridophytes, the gametophytes tend to germinate inside the protective covering of the spore. This process is continued in the seed plants, and to consider first the microspore, the sequence of events in gymnosperms is as follows: the original spore nucleus divides mitotically to give the first vegetative nucleus, and another nucleus which then divides to produce the second vegetative nucleus and an antheridial nucleus. Of these nuclei, the two vegetative nuclei are usually considered to be the vestigial representatives of the prothallus

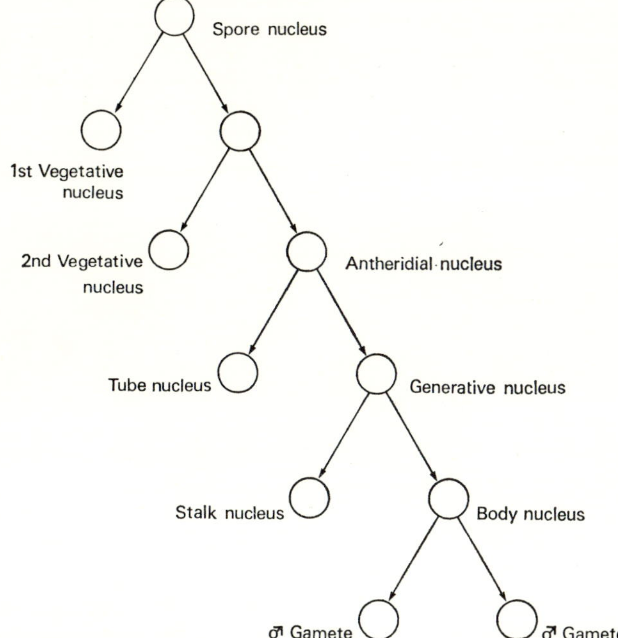

Fig 10.2. Sequence of divisions of microspore nucleus in gymnosperms

or gametophyte, whereas the antheridial nucleus is all that remains of the male reproductive organs. This reduced antheridium then undergoes further division to a tube-nucleus, which is considered to control the functions of the pollen-tube, and the generative nucleus, which itself then divides to give a stalk nucleus and a body nucleus. The latter passes into the pollen-tube to give two male gametes, and of these, one ultimately fertilises the ovum whilst the other nuclei disintegrate.

The pollen-tube itself, as mentioned above, is a new invention of the seed plants. It enables internal fertilisation to occur by providing an aqueous canal from antheridium to the archegonium, through which the male gamete can travel without being exposed to the dehydrating rigours of the terrestrial environment. The sureness of fertilisation is reflected in the small number of antherozoids produced, two per antheridium in gymnosperms, although the risky stage in the life-cycle is emphasised by the excessive numbers of microspores (pollen grains) produced. *Cycas* and *Gingko*, a primitive gymnosperm tree, demonstrate a probable ancestral stage in the evolution of pollination. They produce a short pollen-tube, but instead of this releasing two vestigial male gametes in the form of nuclei, they produce ciliated motile antherozoids, which complete the journey to the ovum under their own power, albeit very slowly.

In angiosperms, the reduction of the male gametophyte is taken a stage further. The original microspore nucleus divides to form a tube nucleus and a generative nucleus. The generative nucleus then later divides once again to form the two gametes. Of these three haploid

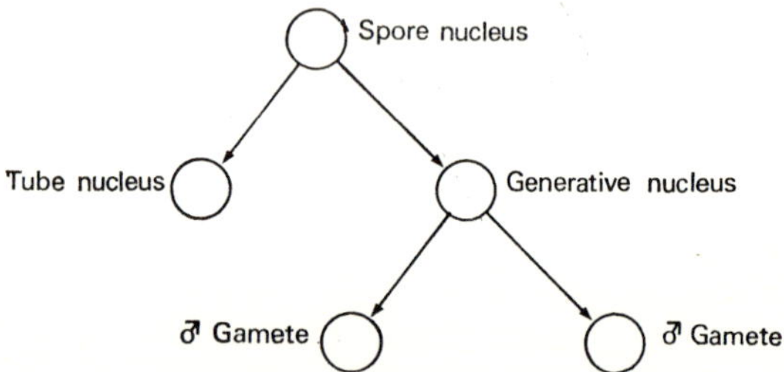

Fig 10.3. Sequence of divisions of microspore nucleus in angiosperms

nuclei, however, only the tube nucleus then disintegrates. One of the gametic nuclei fuses with the ovum, but the other fuses with the endosperm nuclei of the embryosac, to produce another new invention of the angiosperms, namely, triploid endosperm.

The megaspore which remains inside the nucellus germinates to form the female prothallus. In gymnosperms this development is slow and it results in a gametophyte more complex than that of the angiosperm. This again probably represents an intermediate stage in the reduction to the angiosperm condition. The gymnosperm megaspore first undergoes free nuclear division to form a multinucleate female gametophyte within the nucellus. A variable number of archegonia are then produced at the micropylar end of the embryosac, each of which is surrounded by a layer of nutritive cells. The archegonia themselves are each composed of an ovum, a ventral canal cell and a few neck canal cells. The homology here with the bryophyte and pteridophyte archegonia is very clear. Fertilisation then occurs, as described above, but although the ova of the several archegonia in the embryosac may all be fertilised and start to develop, almost invariably only one embryo is present in the mature seed, the rest having disintegrated.

The development of the megaspore in angiosperms is considerably further simplified. The ovule is enclosed within a carpel, but consists as in the gymnosperm, of the megasporangium (nucellus) with its integuments and a stalk or funicle by means of which the ovule is attached to the inside of the ovary. The embryosac with its remaining haploid nucleus then develops as follows. The nucleus divides mitotically and the daughter nuclei formed migrate to opposite ends of the embryosac. They then each undergo two further mitotic divisions, and of the four nuclei produced at each end, one from each group migrates to the middle. The resulting situation is then three nuclei (ovum+ synergids) at the micropylar end; three nuclei (antipodal cells) at the opposite end of the embryosac, and two endosperm nuclei in the middle, which sooner or later fuse. This is the structure of the fully developed embryosac at the moment of fertilisation. The homologies here are not quite as clear as they were in the gymnosperm, but it is generally believed that the ovum and synergids represent the vestigial remains of a single archegonium. Thus the female reproductive organs are finally reduced to one. The triploid fusion nucleus produces the endosperm, which is new to the angiosperms. and not homologous to the endosperm of gymnosperms.

One of the most controversial points in the consideration of the evolution of flowering plants is the derivation of the carpel. The oldest theory, which seems also the most logical, has been attacked many times on many different grounds, but because it explains observed structures very well, it will be expounded here. The theory suggests that the carpel is homologous with the megasporophyll, which instead of opening and displaying the sporangia, remains closed as in the bud, with the edges of the sporophyll fused together, thus protecting the megasporangia, megaspores, and finally the seed. This hypothesis also explains beautifully the structures of more complex, multicarpellary ovaries, which can easily be derived from the fusion together of several carpels in diverse ways.

The stamen also is considered to be a sporophyll, this time bearing the microsporangia, which are embedded in the tissue of the anther.

This sporophyll origin of the carpels and stamens is supported by a group of Mesozoic gymnosperms assigned to the genus *Caytonia*. Here the ovules were borne within an almost closed megasporophyll, and the pollen was produced within pollen-sacs which dehisced in normal angiosperm fashion. The "stamen" lacked a filament, however, and was radially symmetrical, thus removing it from the possible ancestry of angiosperms, but it is quite feasible that the trend demonstrated in *Caytonia* paralleled the development occurring in the group of plants actually ancestral to the angiosperms.

In seed plants, the vagaries of external fertilisation and spore dispersal are replaced by the vagaries of seed and microspore dispersal. The latter process (pollination) must result in the pollen grain alighting on a receptive stigma of the same species of plant. Furthermore, self-pollination is often obviated by a variety of devices, so that the pollen grains may have to travel considerable distances to a comparatively small target. This transport is effected in living plants by the agencies of wind, animals, and water, and the question arises as to which of these three methods is the most primitive. The least economic method is wind dispersal, as the pollen grains here are shed three-dimensionally into the air and may become widely dispersed away from the receptive flowers very rapidly. Water dispersal, which is relatively rare, limits the spread to two dimensions, but animal dispersal is directional and

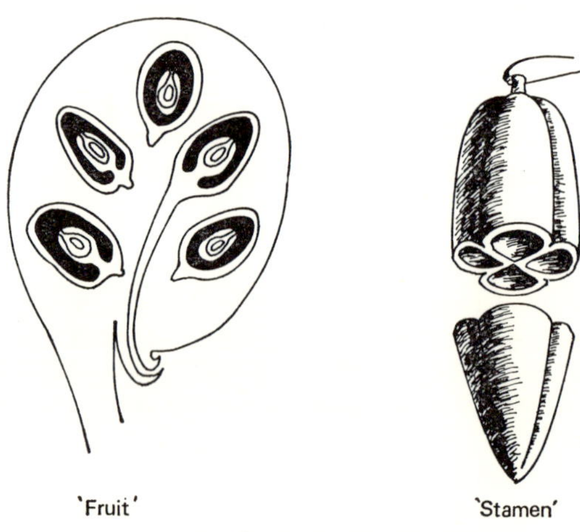

Fig 10.4 ***Caytonia***

therefore much less wasteful of resources. In gymnosperms, the pollen is always dispersed by wind, but is it therefore correct to assume that in angiosperms the earliest members of the group were wind-pollinated? Living anemophilous flowers are all generally more advanced in structure than their insect pollinated relatives, and this suggests that perhaps animal pollination is more primitive among angiosperms. Certainly the colourful petals and powerful scents of the flowers of many primitive angiosperms can only have evolved to attract animals to themselves to facilitate pollination. We have mentioned animals rather than insects, because many of the larger tropical flowers are pollinated by birds and bats, and amongst the insects, not only the hymenoptera and lepidoptera are responsible for pollination, but also the dipterans (some of them attracted by a floral scent of rotting carrion), and coleopterans. The fact that the explosive radiation of angiosperms that occurred during the Cretaceous coincided exactly with a similar radiation in the modern pollinating insect orders confirms that insect pollination is probably more primitive than wind-pollination, at least in the living families of flowering plants. Whether or not bird-pollination preceded insect pollination is an interesting, but far more controversial point.

We have already considered the theory which derives carpels and stamens from specialised spore-bearing leaves, and we have said above that certain flowers are more advanced in structure than certain others. What then are the criteria for deciding on relative degrees of advancement of structure? It is generally agreed that the most primitive floral arrangement found in living angiosperms is represented by the Magnolia family. The primitive features present in these flowers include the generally large size, the lack of distinction between calyx and corolla, the numerous stamens, and the numerous and separate carpels. The different floral parts are also characteristically arranged on the receptacle in a series of spirals. Advancement and specialisation therefore will involve reduction in overall size, reduction in number and fusion of parts, and finally a separation of the male and female flowers to the extent where they may be borne on separate plants.

The Durian Theory

The evolution of the flower, the evolution of pollination mechanisms, and the evolution of the seed, almost complete the story of the advancement of the reproductive mechanisms of angiosperms. The one additional feature which must be mentioned is the development of the fruit. This arises from the ovary wall as an outer cover to the seeds, and in living plants the fruit may become succulent and edible or dry and explosive. Which of these two extremes, however, is closest to the ancestral fruit? The Durian theory, expounded well by Corner, propounds that the most primitive fruits were fleshy and dehiscent and large; very similar in fact to the fruits of the tree belonging to the genus *Durio*, which are found in the tropical rain forest of south-east Asia today. These fruits, the theory maintains, were brightly coloured and contained seeds covered with fleshy coloured arils. An aril is a structure growing round a seed from the funicle originally, which produces, in effect, a third integument. From these fruits were then derived the various existing types of fruit by drying up the fruit wall to form pods, by indehiscence to form berries, or by lignification to give nuts.

This is an interesting theory, but by no means universally accepted, as it is based on observations of living forms in tropical rain forests, which may not represent the cradle of angiosperm evolution. The brightly coloured durians today attract mammals of various kinds, including monkeys, by means of which the seeds are distributed.

There are many other aspects of spermatophyte evolution that are not discussed here. These include the anatomical arrangement of the vascular tissues, the histology of the mechanical tissue, and the different shapes and structures of leaves. These are all interesting study topics, but they require a basic wide knowledge of the variety of plants which takes a long time to acquire and they are more apt to be misleading than floral and reproductive characters, because they are more likely to be specialised adaptations to environmental pressures.

Fungi and Fungal Associations

To omit to mention the evolution of fungi in this account would, according to some modern views, be to omit to mention a third of the categories of living organisms. For it is frequently accepted nowadays that fungi do not form a sub-division of the Thallophyta, but should be elevated to the status of a kingdom, equivalent to that of the plants and animals. The reasons for this splitting are numerous and sound. Firstly, no fungi possess any photosynthetic pigments, and therefore the mode of nutrition is universally heterotrophic, being either saprophytic or parasitic. Fungi are also dissimilar to plants in that their structure is not cellular, but consists of long tubular hyphae with cytoplasmic contents in which are embedded numerous nuclei. The internal division of the hyphae is effected by transverse septa, which are irregularly arranged, and may be completely absent in one class of fungi, the Phycomycetes, of which *Mucor* is a member. Thirdly the hyphal wall is composed of a substance more akin biochemically to chitin than to cellulose.

Several of these reasons propounded for not classifying fungi as plants might appear to place them more close to the animal kingdom, but despite the similarities in nutrition and biochemistry, the lack of any typical animal features such as limited growth and movement preclude this being a serious suggestion of affinities. A further point which is a feature unique to the fungi is that whereas multicellular plants and animals are composed of tissues based on histologically distinct cells, fungal organs are all composed of undifferentiated tubular hyphae. Furthermore, the fact that even when septa are formed they are perforated by a central pore has led some workers to suggest that fungi are essentially uni- or a-cellular and therefore should be classified as Protistans. What does seem clear is that the fungi comprise a sufficiently uniform and distinct group of organisms to warrant consideration as a third kingdom.

That fungi evolved from true photosynthetic plant ancestors is the most probable hypothesis. There is positive circumstantial evidence to support this from amongst living forms. *Clavaria*, for instance, shows an internal arrangement of its hyphae into a false parenchyma very similar to that found in the red algae. Also *Pythium*, a common parasitic and saprophytic fungus causing damping off in seedlings, shows remarkable similarities in both its sexual and asexual reproduction to *Vaucheria*, a filamentous green alga. *Vaucheria* also resembles fungi in generally consisting of filaments which are not internally divided into cells. The fungal habit may thus have developed fairly early on in the history of plants, and has played a vital role ever since, in assisting saprophytically with the breakdown of dead organic matter, thus assisting in both the carbon and nitrogen cycles. Despite the probable early origin of fungi, their main evolutionary radiation did not come for a long time, until the angiosperms were also spreading rapidly. This conclusion can be drawn partly from the nature of the earlier forms of vegetation, which are not readily amenable to fungus attack, and partly from the direct evidence of the remains of the Carboniferous coal-measure plants which obviously were not broken down by fungi after their death.

In addition to the large numbers of species of saprophytic fungi, there are the symbiotic and parasitic forms. Both these modes of life were adopted comparatively late in the history of fungus evolution, as they require a much more subtle adaptation to their living companions than does the saprophytic mode of life. Most of the parasitic forms have as their hosts angiosperms or higher animals. This is an indication of the point in time when the relationship first arose. The symbiotic fungi, however, present more of an evolutionary problem, as symbiosis is often regarded as a perfected form of parasitism, and yet the relatively few fungal symbiotic relationships almost certainly arose before many, if any, parasitic relationships had been effected. Thus what is probably the best known fungal symbiotic situation, that of the lichens, is a very long established one, and may even represent one of the experimental stages in the initial colonisation of the land by plants. Nothing is known, however, of the manner in which the algae and fungi first came together in this way, nor are their mutual contributions fully understood today.

The other symbiotic relationship involving fungi is perhaps easier to understand with respect to its origins, but is almost equally clouded in mystery as to the exact physiological interrelationship between its two members. This is the mycorrhizal association often found between the roots of trees and heathers, and various fungi. The exact nature of the anatomical relationship is variable, hence the classification into ectotrophic and endotrophic forms, and this implies that this symbiosis may have developed from an originally parasitic situation, over which the host has effected some degree of control.

A third fungal relationship, which has been likened to a mycorrhizal association is found in bryophytes. Here, fungal hyphae penetrate the cells of the gametophyte, and apparently the latter does not suffer in any way. This relationship, however, has never yet been thoroughly investigated.

Further Reading

Alexopoulos, C. J. (1952) *Introductory Mycology*. London.

Corner, E. J. H. (1964) *The Life of Plants*. London.

Coulter, J. M. and Chamberlain, C. J. (1917) *Morphology of Gymnosperms*. Chicago.

Duncan, V. K. (1959) *A Guide to the Study of Lichens*. Arbroath.

Garrett, S. D. (1956) *Biology of Root-Infecting Fungi*. Cambridge University Press.

Harley, J. (1959) *The Biology of Mycorrhiza*. London.

Large, E. C. (1940) *The Advance of the Fungi*. London.

McClean, R. C. and Ivimey-Cook, W. R. (1951 and 1967) *Textbook of Theoretical Botany*. Vols. I and III. London.

Sprone, K. R. (1967) *The Morphology of Gymnosperms*. London.

Walton, J. (1953) *An Introduction to the Study of Fossil Plants*. London.

index